LONDON MATHEMATICAL SOCIETY LECTURE NOTE SERIES

Managing Editor: Professor I. M. James, Mathematical Institute, 24-29 St. Giles, Oxford

Prospective authors should contact the editor in the first instance.

Already published in this series

1. General cohomology theory and K-theory, PETER HILTON.
2. Numerical ranges of operators on normed spaces and of elements of normed algebras, F. F. BONSALL and J. DUNCAN.
4. Algebraic topology: A student's guide, J. F. ADAMS.
5. Commutative algebra, J. T. KNIGHT.
8. Integration and harmonic analysis on compact groups, R. E. EDWARDS.
9. Elliptic functions and elliptic curves, PATRICK DU VAL.
10. Numerical ranges II, F. F. BONSALL and J. DUNCAN.
11. New developments in topology, G. SEGAL (ed.).
12. Symposium on complex analysis Canterbury, 1973, J. CLUNIE and W. K. HAYMAN (eds.).
13. Combinatorics, Proceedings of the British combinatorial conference 1973, T. P. McDONOUGH and V. C. MAVRON (eds.).
14. Analytic theory of abelian varieties, H. P. F. SWINNERTON-DYER.
15. An introduction to topological groups, P. J. HIGGINS.
16. Topics in finite groups, TERENCE M. GAGEN.
17. Differentiable germs and catastrophes, THEODOR BRÖCKER and L. LANDER.
18. A geometric approach to homology theory, S. BUONCRISTIANO, C. P. ROURKE and B. J. SANDERSON.
19. Graph theory, coding theory and block designs, P. J. CAMERON and J. H. VAN LINT.
20. Sheaf theory, B. R. TENNISON.
21. Automatic continuity of linear operators, ALLAN M. SINCLAIR.
22. Presentations of groups, D. L. JOHNSON.
23. Parallelisms of complete designs, PETER J. CAMERON.
24. The topology of Stiefel manifolds, I. M. JAMES.
25. Lie groups and compact groups, J. F. PRICE.
26. Transformation groups: Proceedings of the conference in the University of Newcastle upon Tyne, August 1976, CZES KOSNIOWSKI.
27. Skew field constructions, P. M. COHN.
28. Brownian motion, Hardy spaces and bounded mean oscillation, K. E. PETERSEN.
29. Pontryagin duality and the structure of locally compact abelian groups, SIDNEY A. MORRIS.
30. Interaction models, N. L. BIGGS.
31. Continuous crossed products and type III von Neumann algebras, A. VAN DAELE.
32. Uniform algebras and Jensen measures, T. W. GAMELIN.

London Mathematical Society Lecture Note Series 15

An Introduction to Topological Groups

P. J. HIGGINS

Cambridge University Press
Cambridge
London · New York · Melbourne

Published by the Syndics of the Cambridge University Press
The Pitt Building, Trumpington Street, Cambridge CB2 1RP
Bentley House, 200 Euston Road, London NW1 2DB
32 East 57th Street, New York, NY 10022, USA
296 Beaconsfield Parade, Middle Park, Melbourne 3206, Australia

Library of Congress Catalogue Card Number: 74-82222

ISBN 0 521 20527 1

First published 1974
Reprinted 1979

First printed in Great Britain
at the University Press, Cambridge
Reprinted in Great Britain by Redwood Burn Ltd
Trowbridge and Esher

Contents

Preface

In 1969, and again in 1972, I gave an introductory course on topological groups at the University of London. The audience consisted largely of first-year postgraduate students in algebra or number theory and the course was designed to meet their needs, which I took to be a knowledge of basic facts about various general types of topological groups and enough about the Haar integral to enable them to appreciate its use in such contexts as representation theory and algebraic number theory. Some of the students had little or no background in topology and for this reason topological concepts were developed *ab initio*, whereas the rudiments of group theory were assumed.

After the 1969 course there was some demand from new students for copies of the notes and so it was decided to produce a duplicated set of notes of the 1972 lectures. I am extremely grateful to Robert Coates, Michael Rutter and Anthony Solomonides for all their hard work in preparing the mimeographed version on which the present notes are based. I have made a number of minor changes and amplified some passages when this seemed advisable for a wider audience. I have also added an informal section on the representation theory of compact groups which I intended to include in the course but for which there was no time. Its purpose is to illustrate one practical application of the Haar integral and to encourage further reading.

King's College, London
1974

P. J. Higgins

I · Preliminaries

1. Historical notes

Abstract topological groups were first defined by Schreier in 1926, though the idea was implicit in much earlier work on continuous groups of transformations. The subject has its origins in Klein's programme (1872) to study geometries through the transformation groups associated with them, and in Lie's theory of continuous groups arising from the solution of differential equations. The 'classical groups' of geometry (general linear groups, unitary groups, symplectic groups, etc.) are in fact Lie groups, that is, they are analytic manifolds and their group operations are analytic functions. On the other hand, Killing and Cartan showed (1890) that all simple Lie groups are classical groups, apart from a finite number of exceptional groups.

In 1900 Hilbert posed the problem (No. 5 of his famous list) whether every continuous group of transformations of a finite-dimensional real or complex space is a Lie group. The twentieth-century habit of axiomatising everything led to a more abstract formulation of this problem. A topological group is a topological space which is a group with continuous group operations, and the question is: What _topological_ conditions on a topological group will ensure that it has an analytic structure which makes it a Lie group? Since integration was a major tool in the study of Lie groups, especially their representations, it became important to establish the existence of appropriate integrals on general classes of topological groups. This was achieved by Haar in 1933 for locally compact groups with countable open bases. Von Neumann (1934) gave another proof for arbitrary compact groups, making the representation theory of compact Lie groups immediately available for all compact groups and so solving Hilbert's problem in this special case. Haar's method of integration was extended to all locally compact groups by Weil (1940). However, there are serious obstacles to extending the representation theory to locally compact groups, and it was not until 1952 that Hilbert's problem was settled by Gleason, Montgomery and Zippin. Their answer can be formulated as follows: a topological group is a Lie group if and only if it is locally Euclidean; alternatively, it is a Lie group if and only if it is locally compact and does not have arbitrarily small subgroups, that is, the identity element has a compact neighbourhood containing no non-trivial subgroups.

Although the theory of topological groups was developed mainly in order to study groups of Lie type and its impetus came from problems in analysis, it soon proved to be useful also in purely algebraic contexts. Certain algebraic constructions lead to groups having natural topological structures which are somewhat pathological from an analyst's point of view. Examples are power-series rings, Galois groups of infinite field extensions, and p-adic groups. The pathology lies in the existence of arbitrarily small subgroups, but in most important cases the groups are actually locally compact and integration is therefore possible on them. The algebraist must be familiar with these facts and this course is designed to make them available to him. It will, I hope, also serve as an introduction to topological groups and the Haar integral for students of other branches of mathematics. However, the general flavour of the development is more algebraic than is usual in such an introduction, and the use of analytical arguments has been kept to a minimum. The only pre-requisites for the first three chapters are a few facts from elementary group theory. Chapter IV is less rigorous and demands more of the reader.

2. Categories

Definition. A category $\mathcal{C}$ is a structure comprising the following data:

(i) a class whose members A, B, C, ... are called the objects of $\mathcal{C}$;

(ii) for each pair of objects A, B, a set $\mathcal{C}(A, B)$, called the set of morphisms from A to B (in $\mathcal{C}$); we write $f : A \to B$ to mean that $f \in \mathcal{C}(A, B)$;

(iii) for each triple of objects A, B, C, a law of composition

$$\mathcal{C}(A, B) \times \mathcal{C}(B, C) \to \mathcal{C}(A, C);$$

that is, for $f : A \to B$ and $g : B \to C$, there is defined a 'composite' morphism $fg : A \to C$. (Note: we have adopted a right-handed notation for morphisms which is contrary to the current practice of many authors.)

These data are subject to the following axioms:

I. Associative Law. If $f : A \to B$, $g : B \to C$ and $h : C \to D$, then $(fg)h = f(gh)$.

II. Identities. For each object A, there is a morphism $e_A \in \mathcal{C}(A, A)$ such that for all $f : A \to B$, $e_A f = f$, and for all $g : C \to A$, $ge_A = g$.

Examples. (i) The category $\mathcal{S}$ of sets. Its objects $A, B, C, \ldots$ are sets, and the morphisms $f : A \to B$ are functions (maps) from A to B. Composition is ordinary composition of functions. The identities are the maps

$$e_A = \iota_A : a \mapsto a, \text{ for } a \text{ in } A.$$

(Note: in any category in which the morphisms are certain maps between sets, and composition is the ordinary one of maps, associativity follows immediately.)

(ii) The category $\mathcal{G}$ of groups. Its objects are groups, and the morphisms $A \to B$ are group homomorphisms. Composition and identities are as in $\mathcal{S}$. (The point is that if f, g are group homomorphisms, so is fg; and ι_A is always a group homomorphism.)

(iii) The category $\mathcal{T}$ of topological spaces and continuous mappings, defined as follows: Given a set X, a topology $\mathcal{U}$ on X is a collection of subsets of X, called 'open subsets of X', satisfying:

(a) the intersection of two open sets is open (whence so is any finite intersection),

(b) an arbitrary union of open sets is open, and

(c) the empty set $\emptyset$, and X itself, are open subsets of X.

We then call $(X, \mathcal{U})$ a topological space. Notation: if $\mathcal{U}$ is fixed, we shall say simply 'X is a topological space'.

The objects of $\mathcal{T}$ are the topological spaces. Their morphisms are mappings $f : A \to B$ (A, B topological spaces) which are continuous, i.e. such that for all U open in B, the pre-image $Uf^{-1} (= \{x \in A; xf \in U\})$

is open in A. Composition and identities are as in $\mathcal{S}$. (Check: f, g continuous $\Rightarrow$ fg continuous.)

The language of categories enables one to define certain familiar concepts very generally. This is helpful in comparing and contrasting analogous situations in different mathematical contexts.

Definition. Let $\mathcal{C}$ be a category. Then $f : A \to B$ in $\mathcal{C}$ is an isomorphism (in $\mathcal{C}$) if it is invertible in $\mathcal{C}$, i.e. if it has an inverse $g : B \to A$ in $\mathcal{C}$ such that $gf = e_B$, $fg = e_A$. If such a morphism g exists, it is unique, and we write $g = f^{-1}$. We also say that A and B are $\mathcal{C}$-isomorphic and write $A \underset{\mathcal{C}}{\cong} B$.

Examples. The isomorphisms in $\mathcal{S}$ are the bijections (1 - 1 correspondences).

The isomorphisms in $\mathcal{G}$ are the group isomorphisms (bijective group homomorphisms). If a group homomorphism f is bijective, its set-inverse f^{-1} is automatically a group homomorphism.

The isomorphisms in $\mathcal{T}$ are the homeomorphisms, i.e. bijective maps f such that f is continuous and f^{-1} is continuous. There are continuous bijections in $\mathcal{T}$ which are not isomorphisms, i.e. their set-inverses are not continuous.

Salient facts about $\mathcal{S}$. Subsets: If $S \subseteq T$, the 'inclusion' map $j : S \to T$ in $\mathcal{S}$, $(sj = s)$ is an injection and has the following universal property:

if $f : A \to T$ in $\mathcal{S}$, and $Af \subseteq S$, then $\exists!\, f^* : A \to S$ such that $f = f^*j$.

Pictorially:

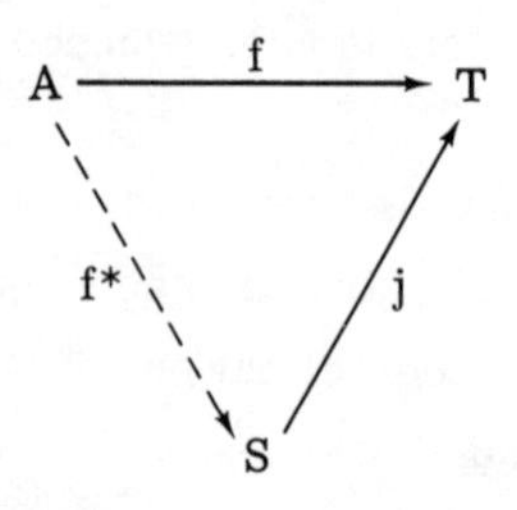

$\exists!\, f^*$ such that the diagram commutes.

($\exists!$ means 'there exists a unique...')

Quotient sets: If A, B are sets, a correspondence from A to B is a subset γ of $A \times B$. If also δ is a correspondence from B to C, $\delta \subseteq B \times C$, we obtain a correspondence $\gamma\delta \subseteq A \times C$ by letting $(a, c) \in \gamma\delta$ if and only if $\exists b \in B$ such that $(a, b) \in \gamma$, $(b, c) \in \delta$. We also define $\gamma^{-1} \subseteq B \times A$ by the rule: $(b, a) \in \gamma^{-1}$ if and only if $(a, b) \in \gamma$. The identity correspondence $\iota_A : A \to A$ is $\{(a, a);\ a \in A\}$. In this notation a function $f : A \to B$ is a correspondence such that $ff^{-1} \supseteq \iota_A$, and $f^{-1}f \subseteq \iota_B$.

An equivalence relation on a set A is a correspondence $\gamma \subseteq A \times A$ such that $\gamma \supseteq \iota_A$ (γ is reflexive), $\gamma = \gamma^{-1}$ (γ is symmetric) and $\gamma\gamma \subseteq \gamma$ (γ is transitive). Given an equivalence relation ρ on A, A is partitioned into equivalence classes $(a) = \{b \in A;\ (a, b) \in \rho\}$. The quotient A/ρ is the set of equivalence classes, and there is a canonical map $q : A \to A/\rho$, $a \mapsto (a)$. This q is a surjection ('an onto map').

For $f : A \to B$, ff^{-1} is an equivalence relation on A whose classes are the 'fibres' of f, namely the sets of form bf^{-1}, for $b \in Af$

Definition. ff^{-1} is the kernel of f, denoted by $\mathrm{Ker}(f)$.

The quotient map $q : A \to A/\rho$, where ρ is an equivalence relation, has the following universal property: if $f : A \to B$ and $\rho \subseteq \mathrm{Ker}(f) = ff^{-1}$, $\exists!\ f^* : A/\rho \to B$ such that $qf^* = f$.

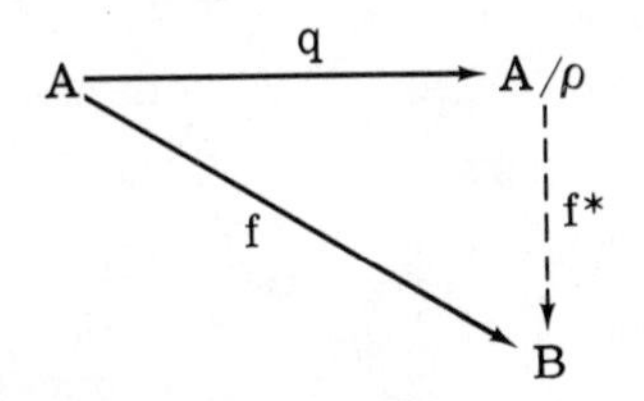

$\exists!\ f^*$ such that the diagram commutes.

Proposition 1 ($\mathcal{S}$). (First isomorphism theorem.) Let $f : A \to B$ be any map. Let ρ be its kernel with quotient map $q : A \to A/\rho$. Let

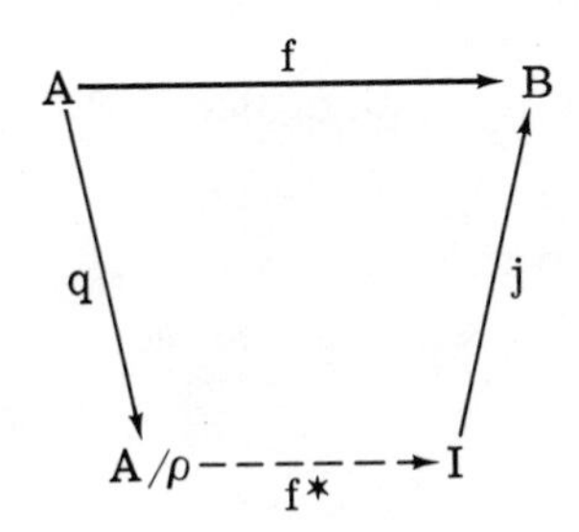

$I = Af \subseteq B$, with inclusion $j : I \to B$. Then (by the universal properties for quotients and subsets), $\exists!\ f^* : A/\rho \to I$ such that $qf^*j = f$. Further, f^* is an isomorphism in $\mathcal{S}$.

Products in $\mathcal{S}$: If I, S are any sets and $x : I \to S$ any map, we often speak of x as a family (indexed set) of elements in S. We use the notation $\{x_i\}_{i\in I}$ for the family, where $x_i = ix$, and we call I the indexing set of the family. For example a sequence of real numbers $(x_n)_{n\in\mathbf{N}}$ is a function $x : \mathbf{N} \to \mathbf{R}$. If $\{A_i\}_{i\in I}$ is a family of sets, we define:

$$\prod_{i\in I} A_i = \{\{a_i\}_{i\in I};\ \forall\, i \in I,\ a_i \in A_i\}.$$

If I is finite, $I = \{1, \ldots, n\}$, then $\prod_{i\in I} A_i$ is $\mathcal{S}$-isomorphic with $A_1 \times A_2 \times \ldots \times A_n$.

Suppose that $a = \{a_i\}_{i\in I}$ is a member of $A = \prod_{i\in I} A_i$. We call a_i the ith coordinate of a, and $\pi_i : A \to A_i$, $a \mapsto a_i$ is called the ith coordinate projection. These projections have a universal property: if for each $i \in I$ we are given a map $f_i : B \to A_i$, then $\exists!\ f : B \to A$ such that $f\pi_i = f_i$ for each i. (This map is given by $bf = \{bf_i\}_{i\in I}$.)

We are thus led to the categorical concept of product. Let $\mathcal{C}$ be a category. Suppose that I is a set, and for each $i \in I$ we are given an object A_i of $\mathcal{C}$. Suppose that A is also an object of $\mathcal{C}$, with morphisms $\pi_i : A \to A_i$ for $i \in I$. We say that A is a product in $\mathcal{C}$ of the A_i, with projections π_i, if for every family of morphisms $\{f_i\}_{i\in I}$, with $f_i : B \to A_i$ in $\mathcal{C}$, $\exists!\ f : B \to A$ in $\mathcal{C}$, such that $f\pi_i = f_i$ for every $i \in I$.

Such a product need not exist. However, any two such products must be isomorphic in $\mathcal{C}$ by the 'standard argument' for universal properties, which we write out in full here (it will be left as an easy exercise at all future occurrences).

If A is a product of the objects A_i, with projections π_i, let us (temporarily) write $f_i, \pi_i \rightsquigarrow f$ to mean that the morphisms $f_i : B \to A_i$ determine (uniquely) the morphism $f : B \to A$ satisfying $f\pi_i = f_i$ for all $i \in I$. If A' is another product of the A_i, with projections $\pi'_i : A' \to A_i$, let

$$\pi_i, \pi'_i \rightsquigarrow \alpha \text{ and } \pi'_i, \pi_i \rightsquigarrow \beta .$$

Then $\alpha : A \to A'$ satisfies $\alpha\pi'_i = \pi_i$, and $\beta : A' \to A$ satisfies $\beta\pi_i = \pi'_i$. It follows that $\alpha\beta\pi_i = \pi_i$, for all $i \in I$, and hence

$$\pi_i, \pi_i \rightsquigarrow \alpha\beta .$$

But clearly, $\pi_i, \pi_i \rightsquigarrow \iota_A$, and it follows from the uniqueness clause that $\alpha\beta = \iota_A$. Similarly $\beta\alpha = \iota_{A'}$. Thus $A \underset{\mathcal{C}}{\cong} A'$ and since $\alpha\pi'_i = \pi_i$ and $\beta\pi_i = \pi'_i$, the isomorphism respects the projections.

We may now say that $\mathcal{S}$ has products (i. e. any family of sets has a product in $\mathcal{S}$).

We shall always assume the Axiom of Choice: 'If the sets A_i $(i \in I)$ are non-empty, their product is also non-empty.' This is equivalent (given the other axioms of standard set theory) to Zorn's Lemma.

Definition. A partially ordered set is a set S together with a correspondence $\gamma \subseteq S \times S$ such that: $\gamma \supseteq \iota_S$ (γ is reflexive), $\gamma\gamma \subseteq \gamma$ (γ is transitive), $\gamma \cap \gamma^{-1} \subseteq \iota_S$ (γ is anti-symmetric). We usually write $a \leq b$ to mean $(a, b) \in \gamma$.

Let S denote a fixed partially ordered set.

Definition. $T \subseteq S$ is a chain if $\forall t, t' \in T$, either $t \leq t'$ or $t' \leq t$ (i. e. T is totally ordered by the induced relation $\gamma \cap T^2$.)

Definition. $x \in S$ is maximal in S if $\forall y \in S$, $y \geq x \Rightarrow y = x$.

Definition. $T \subseteq S$ is bounded above in S if $\exists s \in S$ such that $\forall t \in T$, $t \leq s$.

Definition. If every chain in S is bounded above in S, we say that S is inductively ordered.

Zorn's Lemma asserts: 'Every inductively ordered set has a maximal element.'

N. B., it is not enough to know that all countable ascending chains

$(s_1 < s_2 \ldots < s_n \ldots)$ are bounded above. E.g., take $S =$ all countable subsets of $\mathbf{R}$, ordered by $\subseteq$. Any countable union of sets in S is also in S, hence countable ascending chains are bounded above in S (by their unions). However, no set in S can be maximal because, $\mathbf{R}$ being uncountable, we can always adjoin one more element.

3. Groups

We state some familiar facts about groups in categorical language.

Subgroups. Let G be a group. Let H be a subset of G with a group structure defined on it; then H is a subgroup of G if and only if the inclusion map $j : H \to G$ is a morphism of groups. There is at most one group structure on a given subset H for which this is the case. Subgroup inclusion $j : H \to G$ has a universal property, as for sets: if $f : L \to G$ is a morphism in $\mathcal{G}$ such that $Lf \subseteq H$, then $\exists!\ f^* : L \to H$ in $\mathcal{G}$ such that $f^*j = f$.

Quotient groups. For a morphism $f : G \to H$ of groups it is usual to define the kernel of f as the normal subgroup $\{x \in G;\ xf = e\}$ of G, where $e = e_H$ denotes the identity element of H. This is in conflict with our earlier definition in $\mathcal{S}$, namely, $\text{Ker } f = ff^{-1}$, but the two are closely related, each determining the other. (If we write $K = \{x \in G;\ xf = e\}$, then K is the equivalence class of ff^{-1} containing e_G. On the other hand, ff^{-1} is just the equivalence relation on G whose classes are all the cosets of K.) There is no danger in using the same name, Ker f, for both concepts - the context will always make it clear which is intended. For any normal subgroup N or F, the cosets of N are the equivalence classes of the equivalence relation ρ defined by $a\rho b \iff ab^{-1} \in N$. It is usual to write G/N for the quotient set G/ρ in this case. If $q : G \to G/N$ is the corresponding quotient map, then there is a unique group structure on G/N such that q is a morphism of groups. We shall always give G/N this structure. The quotient map $q : G \to G/N$ in $\mathcal{G}$ has a universal property, as for sets: if $f : G \to H$ is a morphism in $\mathcal{G}$ such that $N \subseteq \text{Ker } f$ (i.e., $Nf = \{e\}$), then $\exists!\ f^* : G/N \to H$ in $\mathcal{G}$ such that $qf^* = f$.

Proposition 1 ($\mathcal{G}$). (First isomorphism theorem.) For $f : G \to H$ in $\mathcal{G}$, let $K = \text{Ker } f$, and let $I = Gf$ (a subgroup of H). Let $q : G \to G/N$

be the quotient map and $j : I \to H$ the inclusion map. Then $\exists!\ f^* : G/N \to I$ (in $\mathcal{S}$) such that $qf^*j = f$. Furthermore, f^* is a $\mathcal{G}$-isomorphism.

Products in $\mathcal{G}$. Suppose we are given for each $i \in I$ a group A_i with identity e_i. Form $A = \Pi A_i$ in $\mathcal{S}$. Define $e_A = \{e_i\}$, and for $a = \{a_i\}$, $b = \{b_i\} \in A$, define $a^{-1} = \{a_i^{-1}\}$, $ab = \{a_ib_i\}$. (We note that $a_i, b_i \in A_i \Rightarrow a_i^{-1}, a_ib_i \in A_i$.) Then with these operations A is a group with identity e_A, and the projections $\pi_i : A \to A_i$ are group-morphisms.

If now $f_i : B \to A_i$ are group-morphisms, $\exists!\ f : B \to A$ (of sets) such that $f\pi_i = f_i$; it is given by $bf = \{bf\pi_i\} = \{bf_i\}$. One verifies that f is in fact a group-morphism, and hence that A is the product in $\mathcal{G}$ of the A_i.

4. Topological spaces

Examples of topological spaces:

(i) Any metric space X under the distance topology, i.e., $Y \subseteq X$ is open if and only if $\forall y \in Y$, $\exists \delta > 0$ such that the δ-ball with centre y is contained in Y. (E.g. $\mathbf{R}^n$ with the usual metric.)

(ii) Any set X with the discrete topology (all subsets are open; this is the strongest topology on X, where if S_1, S_2 are two topological spaces with underlying set X, we say S_1 is stronger than S_2, or S_2 is weaker than S_1, if the identity: $S_1 \to S_2$ is continuous, and strictly so if in addition the identity: $S_2 \to S_1$ is not continuous). Warning: some authors use the terms weaker and stronger in the opposite sense.

(iii) Any set X with the trivial topology ($\emptyset$, X are the only open sets; this is the weakest topology on X).

If X is a space with the discrete topology, or Y is a space with the trivial topology, then any mapping $f : X \to Y$ is continuous.

Subspaces. Let X be a topological space and $Y \subseteq X$ a subset. Consider all sets of the form $A \cap Y$, with A open in X. These may be taken as the open subsets of a topology on Y, the subspace (or induced) topology on Y. It is the weakest topology such that the inclusion map $j : Y \to X$ is continuous. 'Subspace' will always mean 'subset with the induced topology'.

Example. The topological n-sphere S^n is defined to be the subset $\{x; \sum_{i=1}^{n+1} x_i^2 = 1\}$ of $\mathbf{R}^{n+1}$ with the subspace topology, $\mathbf{R}^{n+1}$ having the usual metric topology.

Let Y be a subspace of X. Then $j : Y \to X$ has the usual universal property: if $f : Z \to X$ in $\mathcal{T}$ and $Zf \subseteq Y$, then $\exists!\ f^* : Z \to Y$ in $\mathcal{T}$ such that $f^*j = f$. (For we may construct f^* in $\mathcal{S}$, and continuity follows from the observation that, if A is open in X, then $(A \cap Y)f^{*-1} = (Aj^{-1})f^{*-1} = Af^{-1}$ is open in Z.)

Exercise. The subspace topology on Y is the only one for which j has this property (standard argument; Y is then determined up to $\mathcal{T}$-isomorphism).

Note the contrast with $\mathcal{G}$; a subset of a group G need not possess any group-structure making j a group-morphism, but if one exists, it is unique; whereas a subset of a space X always has a topology (which is not unique) making j continuous. Similar remarks apply to quotients in $\mathcal{T}$ which we now define.

Quotient spaces. Let ρ be an equivalence relation on a topological space X, and let $q : X \to X/\rho$ be the quotient map. The quotient (or identification) topology on X/ρ is the strongest one for which q is continuous, i.e., $A \subseteq X/\rho$ is open if and only if Aq^{-1} is open in X. By 'quotient (or identification) space' we shall always mean 'quotient set with the quotient topology', and X/ρ will always denote this space. In this situation, the quotient map q has the universal property that if $f : X \to Z$ is continuous and $\rho \subseteq \text{Ker } f$, then $\exists!\ f^* : X/\rho \to Z$ in $\mathcal{T}$ such that $qf^* = f$. The quotient topology is the only topology on X/ρ such that q has this property. (The proof of these facts is left as an exercise.)

Example. On $\mathbf{R}$ with the usual topology, define ρ by the rule: $(a, b) \in \rho \Leftrightarrow a - b \in \mathbf{Z}$. Then $\mathbf{R}/\rho = \mathbf{R}/\mathbf{Z}$ is a group and, with the quotient topology, is homeomorphic to the subspace S^1 of $\mathbf{R}^2$, via the canonical bijection $\mathbf{Z} + t \mapsto (\cos 2\pi t, \sin 2\pi t)$. (Exercise: Show that this is a homeomorphism.)

N. B. The first isomorphism theorem is false in $\mathcal{T}$. For example, the identity map $S_1 \to S_2$, where S_1 is a stronger topology than S_2 over the same set X, is a continuous bijection but not a $\mathcal{T}$-isomorphism. Thus some care is required in this last exercise.

We will shortly prove a modified form of the first isomorphism theorem. (See Proposition 1 ($\mathcal{T}$) below.)

Definition. If X is a topological space, $Y \subseteq X$ is closed if its complement in X is open. Closed sets satisfy the following properties:

X, $\emptyset$ are closed;

if A, B are closed, so is $A \cup B$;

if A_i is closed for each $i \in I$, so is $\bigcap_{i\in I} A_i$.

The above may be taken as alternative axioms for a topology on X in terms of closed sets. Then $f : X \to Y$ is continuous if and only if for all $B \subseteq Y$, B closed in $Y \Rightarrow Bf^{-1}$ closed in X.

Definition. Let X, Y be spaces. A map $f : X \to Y$ is said to be closed (respectively open) if for all $A \subseteq X$, A closed $\Rightarrow$ Af closed in Y (respectively A open $\Rightarrow$ Af open in Y).

Exercises. (i) The projection $\mathbf{R}^2 \to \mathbf{R}$, $(x, y) \mapsto x$, is open but not closed.

(ii) The injection $\mathbf{R} \to \mathbf{R}^2$, $x \mapsto (x, 0)$, is closed but not open.

(iii) An isomorphism in $\mathcal{T}$ is open and closed.

(iv) If a continuous bijection is open (or closed) it is a $\mathcal{T}$-isomorphism.

(v) If a continuous surjection $f : X \to Y$ is open (or closed) it is a quotient map (that is, we can find an equivalence ρ on X and a homeomorphism $h : X/\rho \to Y$ such that the following diagram commutes:

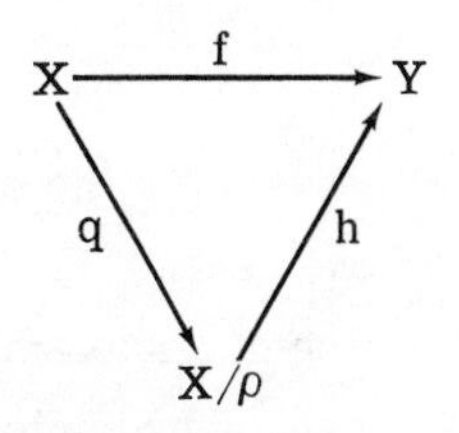

where q is the 'genuine' quotient map $X \to X/\rho$).

(vi) If a continuous injection $f : Z \to X$ is open (or closed), it is an inclusion map (that is, $\exists\, T \subseteq X$ and a homeomorphism $h : Z \to T$, where T has the subspace topology, such that the following diagram commutes:

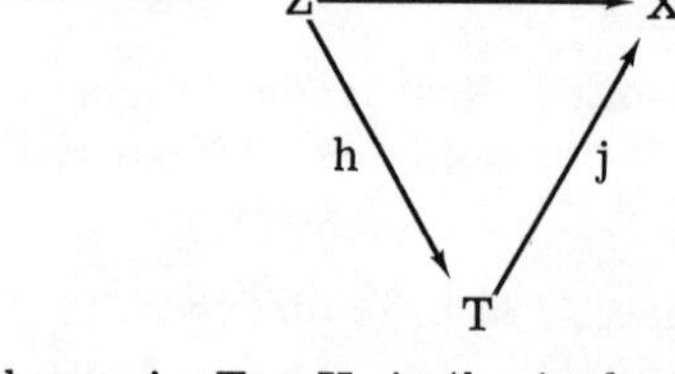

where $j : T \to X$ is the inclusion of T as a subspace in X.

Note: Quotient maps and inclusion maps are not themselves necessarily open (or closed).

Proposition 1 ($\mathcal{T}$). Let $f : X \to Y$ be a morphism in $\mathcal{T}$. Let $\rho = \text{Ker } f\ (= ff^{-1})$, let $q : X \to X/\rho$ be the quotient map (where X/ρ has the quotient topology), and let $Z = Xf$ (as subspace of Y), with $j : Z \to Y$ the inclusion map. Then $\exists!\ f^* : X/\rho \to Z$ in $\mathcal{T}$ such that $qf^*j = f$. If f is open (or closed) then f^* is a $\mathcal{T}$-isomorphism.

Proof. The existence and uniqueness of f^* in $\mathcal{T}$ follow from the universal properties of subspaces and quotient spaces. It is clearly a bijection.

Suppose now that f is open. Let A be open in X/ρ, i.e., let Aq^{-1} be open in X. Then $Aq^{-1}f$ is open in Y, and so $Aq^{-1}fj^{-1}$ is open in Z (because j is continuous). But $qf^*j = f$, so $fj^{-1} = qf^*$, and $Aq^{-1}qf^* = Af^*$ is open in Z. Hence f^* is open and is a continuous bijection, so it is a homeomorphism.

If f is closed, we show in a similar way that f^* is closed, hence a homeomorphism.

(We use the facts that $Sjj^{-1} = S$, $Tq^{-1}q = T$, for all $S \subseteq Z$, $T \subseteq X/\rho$.)

Bases for topologies

Definition. If X is a topological space, a basis for the open sets of X is a family $\{B_i\}_{i \in I}$, where each B_i is an open subset of X, such that any open subset of X can be written as the union of a subfamily, say $\bigcup_{i \in J} B_i$, for some $J \subseteq I$. (If J is empty, we define the above union to be the empty set.)

Examples. (i) In a metric space, all open balls; (ii) in $\mathbf{R}^2$, all open rectangles; (iii) in $\mathbf{R}^2$, all open discs with rational centres, and radii of the form $1/n$, $n \in \mathbf{Z}$.

Note: If X, Y are spaces and $\{B_i\}$ is a basis for the open sets of X, a map $f : Y \to X$ is continuous if and only if every $B_i f^{-1}$ is open in Y.

We can also define a basis for the closed sets of X to be a family $\{D_i\}$ of closed subsets such that any closed set can be written as $\bigcap_{i \in J} D_i$ for some $J \subseteq I$. (Here, if $J = \emptyset$, we define the intersection to be the whole space X.)

If $\{B_i\}_{i \in I}$ is a family of subsets of X, there is a topology on X with the B_i as open basis if and only if the following 'Basis Conditions' are satisfied:

B1 $\bigcup_{i \in I} B_i = X$;

B2 $\forall i, j \in I$, $\exists K \subseteq I$ such that $B_i \cap B_j = \bigcup_{k \in K} B_k$,

and then the topology has precisely all sets of the form $\bigcup_{i \in J} B_i$, for $J \subseteq I$, as open sets.

Let $\{C_\alpha\}_{\alpha \in A}$ be an arbitrary family of subsets of X. Consider all sets of the form

$$B = C_{\alpha(1)} \cap \ldots \cap C_{\alpha(n)}, \quad \text{with } \alpha(j) \in A,$$

i.e., all finite intersections of the sets C_α, where we interpret the empty intersection to be X. Then clearly these sets satisfy **B1** and **B2** above, and so form a basis for a topology $\mathcal{U}$ on X. We say that the sets C_α form a sub-basis for $\mathcal{U}$. It is clear that $\mathcal{U}$ is the weakest topology on X such that all the C_α are open. If X has the above topology $\mathcal{U}$, a map $f : Y \to X$ (Y a topological space) is continuous if and only if every $C_\alpha f^{-1}$ is open.

Products in $\mathcal{T}$. Suppose that $\{X_i\}_{i \in I}$ is a family of topological spaces. Let $X = \Pi X_i$ in $\mathcal{S}$, with projections $\pi_i : X \to X_i$. Let $\mathcal{U}$ be the topology on X with sub-basis all sets of the form $Y_i \pi_i^{-1}$, with Y_i open in X_i; then clearly $\mathcal{U}$ makes all the π_i continuous; it is in fact the weakest topology for which this is so.

Now let $f_i : Z \to X_i$ be continuous maps. Then $\exists!$ map $f : Z \to X$ such that for all $i \in I$, $f\pi_i = f_i$. For each set $Y_i\pi_i^{-1}$ of the sub-basis, we have $(Y_i\pi_i^{-1})f^{-1} = Y_if_i^{-1}$, which is open in Z (because f_i is continuous). It follows that f is continuous. This shows that in the category $\mathcal{T}$, $(X, \mathcal{U})$ is a product of the given objects X_i. Since products in $\mathcal{T}$ are unique (up to $\mathcal{T}$-isomorphism) we see that $\mathcal{U}$ is the only topology on the product set which has this universal property in $\mathcal{T}$. We call this topology $\mathcal{U}$ the <u>product (or Tychonoff) topology on X.</u> A basis for the open sets of $\mathcal{U}$ is obtained by taking all finite intersections of sets $Y_i\pi_i^{-1}$, with Y_i open in X_i. Now $Y_i\pi_i^{-1} = \prod_{j\in J} Y_j'$, where $Y_i' = Y_i$ and $Y_j' = X_j$ for $j \neq i$. Thus a finite intersection of such sets is precisely a set of the form $T = \prod_{i\in I} T_i$, where each T_i is an open subset of X_i and, for all but a finite number of i, $T_i = X_i$. The open sets of the product topology $\mathcal{U}$ are therefore all unions of sets T as above.

Examples. (i) $\mathbf{R}^n$ is the product in $\mathcal{T}$ of n copies of $\mathbf{R}$.

(ii) **Definition.** The <u>n-dimensional torus</u> T^n is the product $\underbrace{S^1 \times S^1 \times \ldots \times S^1}_{n \text{ copies}}$ in $\mathcal{T}$. In particular, T^0 is a single point; $T^1 = S^1$; T^2 is the surface of a doughnut. In general T^n can be parametrized by n complex numbers $(z_1, \ldots, z_n)$ such that $|z_i| = 1$ and T^n is a subspace of $\mathbf{C}^n$. (See Exercise (iv) below.)

<u>Note:</u> If $\{X_i\}_{i\in I}$ is a family of discrete spaces then $\prod_{i\in I} X_i$ is not in general discrete. E.g., consider a countable number of discrete spaces X_i, $i \in \mathbf{N}$, each having two elements. All open sets of $X = \prod_{i\in\mathbf{N}} X_i$ have an infinite number of elements, so X is certainly not discrete. (In fact X is homeomorphic to the Cantor set.) For a more general result see Exercise (v) below.

Exercises. (i) If $X = \prod_{i\in I} X_i$ in $\mathcal{T}$, then the projections $\pi_i : X \to X_i$ are open maps and each X_i is (isomorphic to) a quotient space of X under the projection map π_i.

Definition. In any category $\mathcal{C}$ if $X = \prod_{i \in I} X_i$ and $Y = \prod_{i \in I} Y_i$ with projections $\xi_i : X \to X_i$ and $\eta_i : Y \to Y_i$ respectively and if $f_i : X_i \to Y_i$, then the morphisms $\xi_i f_i : X \to Y_i$ define a unique $f : X \to Y$ such that $\xi_i f_i = f\eta_i$ for all $i \in I$. This f will be denoted by $\prod_{i \in I} f_i$ and will be called the product of the morphisms f_i.

(ii) If the f_i are as above in some category $\mathcal{C}$ and if in addition we have morphisms $g_i : Y_i \to Z_i$ in $\mathcal{C}$, then $\prod_{i \in I} (f_i g_i) = (\prod_{i \in I} f_i)(\prod_{i \in I} g_i)$.

(iii) If I is finite, and the morphisms $f_i : X_i \to Y_i$ in $\mathcal{T}$ are open, then $\prod_{i \in I} f_i$ is open.

<u>Warning:</u> If $h_i : Z \to X_i$ are open in $\mathcal{T}$ then it is not true in general that the induced map $h : Z \to \prod_{i \in I} X_i$ is open, even when I is finite.

(iv) If X_i is a subspace of Y_i for each i then ΠX_i is a subspace of ΠY_i (i.e. the product of inclusion maps is an inclusion map).

<u>Warning:</u> If Y_i is a quotient space of X_i with quotient map $q_i : X_i \to Y_i$ then the map $q = \prod_{i \in I} q_i : \Pi X_i \to \Pi Y_i$ is not in general a quotient map. A counterexample, due to Dieudonné, is as follows: Let $\mathbf{Q}$ be the rational numbers considered as a subspace of $\mathbf{R}$ and let Y be the quotient space obtained by identifying all the integers in $\mathbf{Q}$. If $q : \mathbf{Q} \to Y$ is the resulting quotient map and ι is the identity map on $\mathbf{Q}$ then $\iota \times q : \mathbf{Q} \times \mathbf{Q} \to \mathbf{Q} \times Y$ is not a quotient map.

(v) If X_i $(i \in I)$ are discrete spaces then ΠX_i is discrete if and only if at most a finite number of the X_i have more than one point.

II · Topological groups

1. The category of topological groups

Definition. A topological group is a set G with two structures:

(i) G is a group with respect to $\cdot$, $^{-1}$, e,

(ii) G is a topological space,

such that the two structures are compatible i.e., the multiplication map $\mu : G \times G \to G$ and the inversion map $\nu : G \to G$ are both continuous. In this definition the set $G \times G$ carries the product topology.

Examples. (i) $\mathbf{R}$ regarded as an additive group with the usual metric topology.

(ii) $\mathbf{R}^*$, the set of all non-zero real numbers, regarded as a group with respect to multiplication, with the usual topology.

(iii) Any abstract group is a topological group with respect to the discrete topology.

(iv) $\mathbf{C}^*$, the set of non-zero complex numbers, as a multiplicative group, with the usual topology.

(v) $S^1 \subseteq \mathbf{C}$, $S^1 = \{z : |z| = 1\}$, is a topological group with respect to multiplication and the subspace topology. More generally we have:

(vi) If G is any topological group and H is a subgroup, then H is a topological group with respect to the subspace topology. (Note: (a) $\mathbf{Z}$, the group of integers, is discrete as a topological subgroup of $\mathbf{R}$. (b) $\mathbf{Q}$, the group of rationals, is not discrete as a topological subgroup of $\mathbf{R}$.)

(vii) The general linear group $GL_n(\mathbf{R})$ (the set of all non-singular $n \times n$ matrices with coefficients in $\mathbf{R}$) is a group with respect to multiplication and has a topology induced by the inclusion in $\mathbf{R}^{n^2}$. Operations of multiplication and inversion are given by n^2 rational functions of $2n^2$ and n^2 real variables, respectively, and are continuous. Thus $GL_n(\mathbf{R})$ is a topological group.

(viii) $GL_n(\mathbf{C})$ is a topological group in a similar way.

(ix) The algebra $\mathbf{H}$ of quaternions is a vector space over $\mathbf{R}$ with basis 1, i, j, k. It is trivially a topological group with respect to +, -, 0, isomorphic with $\mathbf{R}^4$. Multiplication of quaternions is given by $i^2 = j^2 = k^2 = -1$, $ij = -ji = k$, $jk = -kj = i$, $ki = -ik = j$, and bilinearity, the basis element 1 acting as identity element. If $h = \alpha 1 + \beta i + \gamma j + \delta k \in \mathbf{H}$, we define the conjugate of h to be $\bar{h} = \alpha 1 - \beta i - \gamma j - \delta k$, and the norm of h to be $|h| = (\alpha^2 + \beta^2 + \gamma^2 + \delta^2)^{\frac{1}{2}}$. Since $h\bar{h} = |h|^2 1$, we see that every non-zero quaternion h has a multiplicative inverse $h^{-1} = |h|^{-2}\bar{h}$. The fact that multiplication is associative can be checked directly on the basis elements. Alternatively, if, for each quaternion $h = \alpha 1 + \beta i + \gamma j + \delta k$ we form the complex matrix $M(h) = \begin{pmatrix} \alpha+\beta i & \gamma+\delta i \\ -\gamma+\delta i & \alpha-\beta i \end{pmatrix}$, we find that $M(h_1 h_2) = M(h_1)M(h_2)$, and associativity in $\mathbf{H}$ follows from associativity of matrix multiplication. The outcome of this is that $\mathbf{H}^*$, the set of non-zero quaternions is a multiplicative group. It also has a topology inherited from $\mathbf{H} \cong \mathbf{R}^4$ and it is a topological group because multiplication and inversion are given by formulae involving only rational functions of the coordinates.

(x) The 3-sphere S^3 is embedded in $\mathbf{H}^*$ as the subgroup of all quaternions of unit norm. Hence S^3 is a topological group with respect to quaternion multiplication.

(xi) Let F be a field and let $| \ |$ be a valuation on F, i.e., a function from F to the set of non-negative real numbers satisfying:

(a) $|x| = 0 \iff x = 0$,

(b) $|xy| = |x||y|$,

(c) $|x + y| \leq |x| + |y|$,

for all $x, y, z \in F$. If we define $d(x, y) = |x - y|$ then, by (a) and (c), d is a metric which induces a topology on F. It is easy to see that F is now a topological group with respect to +, -, 0, and that F^* is a topological group with respect to $\cdot$, $^{-1}$, 1, the proofs being exactly the same as for $\mathbf{R}$. A case of particular importance in number theory is the following. Let $F = \mathbf{Q}$ and let p be a fixed prime number. Every non-zero rational number x can be written in the form $x = p^r m/n$, where $m, n, r \in \mathbf{Z}$ and $p \nmid m$, $p \nmid n$. The integer r is uniquely determined by x. If we define $|x|_p = p^{-r}$ for $x \neq 0$ and $|0|_p = 0$, then $| \ |_p$ is a valuation, called the p-adic valuation, and the resulting topology on $\mathbf{Q}$ is the p-adic topology. Thus every prime p provides two topological groups $(\mathbf{Q}, +)$ and $(\mathbf{Q}^*, \cdot)$ with respect to the p-adic topology.

Note: In the definition of topological group one cannot omit the assumption that inversion is continuous. For example, if we give $\mathbf{Z}$ the topology with open sets $\emptyset$, $\mathbf{Z}$ and all intervals $[n, +\infty)$, then $+$ is continuous but $-$ is not.

Definition. A morphism $f : G \to H$ of topological groups is a continuous group homomorphism.

Clearly topological groups and their morphisms form a category, which we denote by $\mathcal{TG}$.

Note: An isomorphism in $\mathcal{TG}$ is a continuous group homomorphism having an inverse which is also a continuous group homomorphism. It is not true in general that a group isomorphism which is continuous is a $\mathcal{TG}$-isomorphism. For example, let $\mathbf{R}'$ be the reals with the discrete topology. Then the identity map $\mathbf{R}' \to \mathbf{R}$ is a continuous group isomorphism, but its inverse, the identity map $\mathbf{R} \to \mathbf{R}'$ is not continuous.

Examples. (i) $(\mathbf{R}, +) \underset{\mathcal{TG}}{\cong} (\mathbf{R}^{pos}, \cdot)$, the isomorphism being given by $x \mapsto \exp x$ from $\mathbf{R}$ to $\mathbf{R}^{pos}$ and $x \mapsto \log x$ for its inverse.

(ii) The orthogonal group $O_2(\mathbf{R}) = \{ \begin{pmatrix} \cos\theta & \sin\theta \\ -\sin\theta & \cos\theta \end{pmatrix} : \theta \in \mathbf{R} \}$ is a subgroup of $GL_2(\mathbf{R})$. There is a $\mathcal{TG}$-isomorphism $O_2(\mathbf{R}) \to S^1 \subseteq \mathbf{C}^*$ given by $\begin{pmatrix} x & y \\ -y & x \end{pmatrix} \mapsto x + iy$.

2. Subgroups and quotient groups

Any subgroup H of a topological group G is a topological group with respect to the subspace topology, and the inclusion map $j : H \to G$ is a morphism of $\mathcal{TG}$. This morphism has the universal property: given any $f : L \to G$ in $\mathcal{TG}$ with $Lf \subseteq H$, $\exists!\ f^* : L \to H$ in $\mathcal{TG}$ such that $f^*j = f$.

A basic principle: Let G be a topological group and let g be a fixed element of G. The constant map $x \mapsto g$ and the identity map $x \mapsto x$ are continuous maps from G to G, so they induce a continuous map $x \mapsto (g, x)$ from G to $G \times G$. Composing this with the continuous multiplication $G \times G \to G$ we get a continuous map $l_g : x \mapsto gx$ from G to G, called left multiplication (or left translation) by g. This map has inverse $l_{g^{-1}}$ which is also continuous, so l_g is a homeomorphism $G \to G$ (an isomorphism in $\mathcal{T}$, but not in $\mathcal{T}\mathcal{G}$). Similarly all right translations $r_g : x \mapsto xg$ are homeomorphisms $G \to G$. As a consequence G must be a homogeneous space, that is, given $a, b \in G$ there is a homeomorphism $G \to G$ sending a to b. ($l_{ba^{-1}}$ or $r_{a^{-1}b}$ will do.) Thus G looks topologically the same at all points. We can now use translations to transfer topological information from one point to another in any topological group, and this basic method is used in almost every proof in the subject.

Note: Since not all topological spaces are homogeneous it is not always possible to find a group structure on a given topological space which will make it into a topological group.

Notation. If $A, B \subseteq G$, and $g \in G$, where G is a topological group, then

(i) $Ag = Ar_g = \{ag;\ a \in A\}$; Ag is called the right translate of A by g;

(ii) $gA = Al_g$;

(iii) $AB = \bigcup_{b \in B} Ab = \bigcup_{a \in A} aB$;

(iv) $A^{-1} = \{a^{-1};\ a \in A\}$.

Proposition 0. Let G be a topological group $A, B \subseteq G$, $g \in G$. Then

(i) G is a homogeneous space;

(ii) A open implies Ag and gA open;

(iii) A closed implies Ag and gA closed;

(iv) A open implies AB and BA open;

(v) A closed and B finite implies AB and BA closed.

Proof. (i): already proved.

(ii), (iii): l_g, r_g, being homeomorphisms, are both open and closed.

(iv): $AB = \bigcup_b Ar_b$ is a union of open sets and hence open. Similarly for BA.

(v) AB and BA are each the union of a finite number of closed sets and hence closed.

<u>Quotient groups and coset spaces.</u> Let G be in $\mathcal{TG}$ and let H be a subgroup of G. Let $G/H = \{xH;\ x \in G\}$, the set of left cosets of H. The map $q : G \to G/H$ defined by $x \mapsto xH$ defines a quotient topology on G/H. With this topology G/H is called the <u>left coset space of</u> G <u>with respect to</u> H. The quotient map q <u>is an open map</u>; for if S is an open subset of G, then Sqq^{-1} = the union of all left cosets of H which meet S

= SH, which is open by Proposition 0(iv),

and it follows that Sq is open in G/H by the definition of the quotient topology. Now suppose that H is normal in G. Then G/H has a canonical group structure. If μ and μ' are the multiplications in G and G/H and ν, ν' the inversions in G and G/H respectively then μ', ν' are uniquely defined by the commutativity of the following diagrams in $\mathcal{S}$.

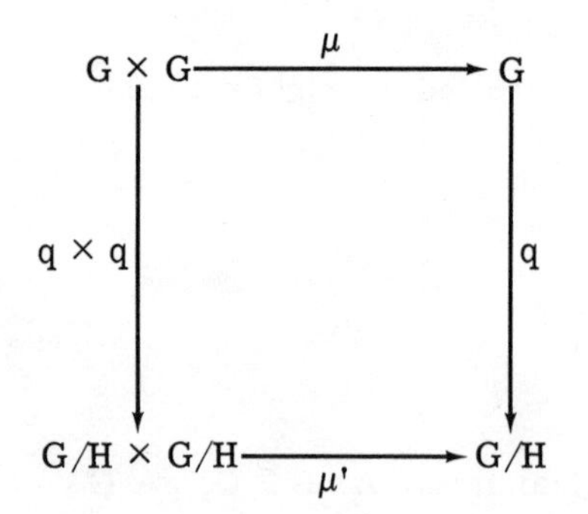

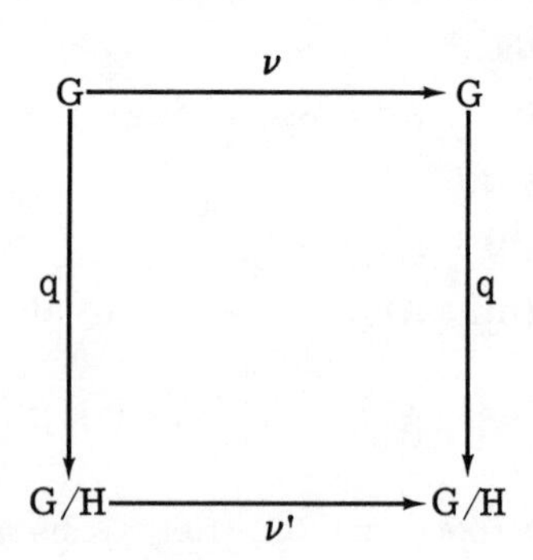

We claim that G/H with this group structure (and the quotient topology) is a topological group, i. e. μ' and ν' are continuous.

Proof. νq is continuous and so by the universal property of q in $\mathcal{T}$ there exists a unique map $\theta : G/H \to G/H$ in $\mathcal{T}$ making $q\theta = \nu q$. However, ν' satisfies the condition $q\nu' = \nu q$ and so $\nu' = \theta \in \mathcal{T}$.

Also μq is continuous and so by the same argument μ' is continuous provided that $q \times q$ is a quotient map. In general it is not the case that the product of quotient maps is a quotient map (see Exercise (iv), p. 15); however, in our case q is <u>open</u> and so $q \times q$ is open (see

Exercise (iii), p. 15). In addition $q \times q$ is continuous and a surjection and hence a quotient map. Thus μ' is continuous and G/H is a topological group.

Proposition 1 ($\mathcal{TG}$). Let G be a topological group, H a subgroup, G/H the left coset space, $q : G \to G/H$ the quotient map. Then:

(i) G/H is homogeneous,

(ii) q is open,

(iii) if $H \lhd G$ (i.e., H is a normal subgroup of G) then G/H is a topological group, with the usual universal property: if $f : G \to L$ in $\mathcal{TG}$ and $Hf = \{e\}$, then $\exists!\ f^* : G/H \to L$ such that $qf^* = f$.

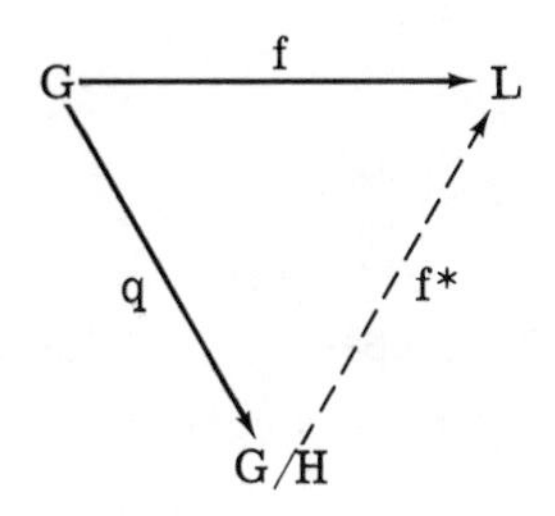

(iv) Let $f : G \to L$ in $\mathcal{TG}$ have kernel K and image $I = Gf \subseteq L$, with quotient map $q : G \to G/K$ and inclusion map $j : I \to L$; then $\exists!\ f^* : G/K \to I$ in $\mathcal{TG}$ such that $qf^*j = f$. The map f^* is always a bijection, and if f is open (or closed) then f^* is a $\mathcal{TG}$-isomorphism.

Proof. (i) is an exercise; (ii) and (iii) have been proved above; (iv) follows from the corresponding results for $\mathcal{G}$ and $\mathcal{T}$.

Examples. (i) Define $f : \mathbf{R} \to S^1$ by $t \mapsto \exp(2\pi it)$. Then f is a morphism from the additive group of reals, with the metric topology, to the multiplicative group S^1 (considered as a subspace of the complex numbers).

The composite map $\mathbf{R} \xrightarrow{f} S^1 \to \mathbf{C} = \mathbf{R} \times \mathbf{R}$ has the continuous components $\cos(2\pi t)$, $\sin(2\pi t)$, hence f is itself continuous. Now $\mathrm{Ker}(f) = \mathbf{Z}$, and $\mathrm{Im}(f) = S^1$. Also f is open. (Proof: the open intervals $(a, b) = I$ $(b > a)$ form a basis for the topology on $\mathbf{R}$. Write $l(I) = b - a$.

If $l(I) > 1$, $f(I) = S^1$ is open in S^1.

If $l(I) = 1$, $f(I) = S^1$ less a point; but $\mathbf{C}$ less a point is open in the whole space, so $f(I)$ is open in S^1.

If $l(I) < 1$, $f(I)$ is an arc without end-points, and so is open in S^1, being the intersection of S^1 with an open disc with the missing end-points on its edge.)

Hence, by Proposition 1(iv), $\mathbf{R}/\mathbf{Z} \cong S^1$ as topological groups.

N. B. f is <u>not</u> closed; e.g., $\{n + (1/n);\ n \text{ integral} > 1\}$ is closed, but $\{f(n+(1/n))\}$ is not closed, as its closure contains $f(0)$.

(ii) In a similar way we have a map $g : \mathbf{C}^* \to \mathbf{R}^{pos}$ in $\mathcal{TG}$, given by $z \mapsto |z|$ which is an open surjection, with kernel S^1, so

$$\mathbf{C}^*/S^1 \cong \mathbf{R}^{pos} \quad \text{in } \mathcal{TG}.$$

(iii) Again, the map $\mathbf{C}^* \to S^1$, given by $z \mapsto z/|z|$ induces a $\mathcal{TG}$-isomorphism $\mathbf{C}^*/\mathbf{R}^{pos} \to S^1$. (See also Exercise (ii) p. 26).

(iv) Considering $M_n(\mathbf{R})$ as real $(n \times n)$-space, the determinant map $\delta : A \mapsto \det(A)$ is given by a polynomial, hence is continuous. By restriction to the subspace of invertible matrices we have a continuous homomorphism of groups

$$\delta : GL_n(\mathbf{R}) \to \mathbf{R}^*,$$

with kernel denoted $SL_n(\mathbf{R})$ (the special linear group). For $x \in \mathbf{R}^*$, the matrix $\mathrm{diag}(x, 1, \ldots, 1)$ has determinant x, so δ is a surjection. We claim that δ is open, and thus

$$GL_n(\mathbf{R})/SL_n(\mathbf{R}) \cong \mathbf{R}^* \quad \text{in } \mathcal{TG}.$$

(<u>Proof that</u> δ <u>is open</u>: Let U be an open set in $GL_n(\mathbf{R})$ and let A be a matrix in U. For real $t \neq 0$, $(tA)\delta = t^n(A\delta) \neq 0$. Since U is open in GL_n (a subspace of $M_n(\mathbf{R})$), there is an open interval J round 1 such that $tA \in U$ for all $t \in J$. Now as t ranges over J, t^n takes every value in some open interval J' round 1, so $t^n(A\delta)$ takes every value in some open interval J'' round $A\delta$. Thus $U\delta \supset J''$ and $U\delta$ is open in $\mathbf{R}^*$.)

(v) In $GL_n(\mathbf{R})$ the set of all 'scalar matrices' tI is a closed normal subgroup, isomorphic as topological group to $\mathbf{R}^*$. The quotient group by this subgroup is (definition) the projective general linear group

$PGL_n(\mathbf{R})$. It is, of course, a topological group with respect to the quotient topology.

Corollary to Proposition 1. Let G be a topological group. Let H, L be subgroups with $H \subseteq L$. Then the two topologies on L/H as (a) quotient space of L and (b) subspace of G/H are the same. Hence, if $H \lhd G$, then every subgroup of G/H is isomorphic in $\mathcal{T}\mathcal{G}$ to a quotient group L/H.

Proof. Let $q : G \to G/H$ be the quotient map and let q' be the induced map $L \to Lq$. By Proposition 1($\mathcal{T}$) it is enough to show that q' is an open map. So suppose that A is an open subset of L. Then $A = L \cap B$ where B is open in G. Since q is an open map (Proposition 1($\mathcal{T}\mathcal{G}$)), Bq is open in G/H, whence $Bq \cap Lq$ is open in Lq. However, since L is a union of left cosets of H, we have $Bq \cap Lq = (B \cap L)q = Aq$, so Aq is open in Lq, as required.

Exercise. Prove that if $H \lhd G$, $L \lhd G$ and $H \subseteq L$, then $(G/H)/(L/H) \cong G/L$ in $\mathcal{T}\mathcal{G}$.

Warning: The analogue in $\mathcal{T}\mathcal{G}$ of the third isomorphism theorem of group theory would say that if A and H are subgroups of the topological group G, with $H \lhd G$, then $A/(A \cap H) \cong (AH)/H$ in $\mathcal{T}\mathcal{G}$. This is false, as the following example shows. Let $G = \mathbf{R}$, $H = \mathbf{Z}$, $A = \lambda\mathbf{Z}$, where λ is irrational. Then $A \cap H = \{0\}$, so $A/(A \cap H) = A$ is isomorphic to $\mathbf{Z}$ with the discrete topology. However $A + H = \mathbf{Z} + \lambda\mathbf{Z}$ is dense in $\mathbf{R}$, i.e., every open set in $\mathbf{R}$ meets it. (Proof: If an additive subgroup $C \neq 0$ of $\mathbf{R}$ has a least positive member d then $C = d\mathbf{Z}$. If it has no least positive member then it contains a strictly decreasing positive sequence $c_1 > c_2 > c_3 > \dots$. Given any interval (p, q), C contains some c with $0 < c < q - p$, namely, $c = c_i - c_{i+1}$ for sufficiently large i. Some multiple nc of this element lies in (p, q), so C is dense. In our case $\mathbf{Z} + \lambda\mathbf{Z}$ is not of the form $d\mathbf{Z}$ since this would imply $1 = dm$, $\lambda = dn$, for some $m, n \in \mathbf{Z}$, whence λ would be rational.) It now follows that $(A + H)/H = (\mathbf{Z} + \lambda\mathbf{Z})/\mathbf{Z}$ is dense in $\mathbf{R}/\mathbf{Z} \cong S^1$. It is isomorphic to $\mathbf{Z}$ as a group, but it is not discrete since every non-empty open set in S^1 meets it in an infinite set, so none of its finite subsets are open.

We briefly describe another notorious example which is perhaps easier to visualise. It concerns subgroups of the two-dimensional torus $T^2 = S^1 \times S^1$ which is a topological group with respect to the product group structure and the product topology (see the next section for products in $\mathcal{TG}$). The function $\phi_\lambda : \mathbf{R} \to T^2$ defined by $t \mapsto (\exp 2\pi i t, \exp 2\pi i \lambda t)$ is a morphism of topological groups. Its image I_λ is a subgroup of T^2 whose nature changes radically according to the value of the real constant λ. If λ is rational, all is well; putting $\lambda = \pm m/n$ where m and n are coprime positive integers we find that $\operatorname{Ker} \phi_\lambda = n\mathbf{Z}$ and I_λ is a topological circle which winds round the torus m times one way and n times the other way. Indeed, the restricted map $\phi^* : \mathbf{R} \to I_\lambda$ is a continuous, open surjection, and therefore $I_\lambda \cong \mathbf{R}/n\mathbf{Z}$, which is $\mathcal{TG}$-isomorphic to S^1. We leave the details as an exercise. If, on the other hand, λ is irrational, then $\operatorname{Ker} \phi_\lambda = \{0\}$ and, as a group, I_λ is isomorphic to $\mathbf{R}$. It winds round the torus infinitely many times, never passing through the same point more than once. However, the topology on I_λ does not match the topology on $\mathbf{R}$ because, although I_λ passes through the identity $\phi_\lambda(0) = (1, 1)$ only once, it passes arbitrarily close to it after winding round the torus suitably often. In fact, every neighbourhood of $\phi_\lambda(0)$ in T^2 contains points $\phi_\lambda(t)$ for an unbounded set of values of t. Hence any neighbourhood of $\phi_\lambda(0)$ in I_λ corresponds to an unbounded subset of $\mathbf{R}$, and the two topologies are quite different. Again we leave the detailed proof of these facts as an exercise. (In order to find points $\phi_\lambda(t)$ close to $\phi_\lambda(0)$ one looks for pairs of integers (s, t) such that $s + \lambda t$ is close to 0. There are plenty of such pairs because, as we showed above, $\mathbf{Z} + \lambda\mathbf{Z}$ is dense in $\mathbf{R}$.)

This example confirms the need for a topological condition in the first isomorphism theorem. It also gives another example showing that the third isomorphism theorem of group theory does not extend to topological groups. We put $G = \mathbf{R} \times \mathbf{R}$, $H = \mathbf{Z} \times \mathbf{Z}$ and consider the line $A = \{(t, \lambda t);\ t \in \mathbf{R}\}$ in G with irrational slope λ. Then $A \cap H = \{(0, 0)\}$ so $A/(A \cap H) \cong A \cong \mathbf{R}$ in $\mathcal{TG}$. However, $(A + H)/H$ is, by the corollary to Proposition 1, a topological subgroup of the quotient group G/H. We shall see in the next section that $G/H \cong T^2$ in $\mathcal{TG}$, the isomorphism being induced by the map $\theta : (x, y) \mapsto (\exp 2\pi ix, \exp 2\pi iy)$ from G to T^2. The subgroup of T^2 corresponding to $(A + H)/H$ is the image of A under θ, which is precisely the subgroup I_λ described above. It follows that $(A + H)/H$ and $A/(A \cap H)$ have different topologies though, of course, they are canonically isomorphic in $\mathcal{G}$.

3. Products

Let $\{G_i\}_{i \in I}$ be a family of topological groups. Their product $G = \Pi G_i$ (as sets) has a natural group structure (product in $\mathcal{G}$) and a natural topology (product in $\mathcal{T}$). With this topology on G, a map $f : X \to G$ is continuous if and only if each $f\pi_i : X \to G_i$ is continuous.

The group operations μ, ν are defined on G so that the following diagrams commute for each i:

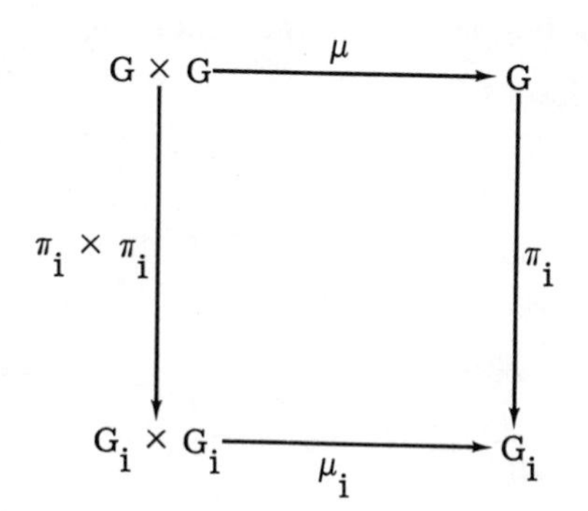

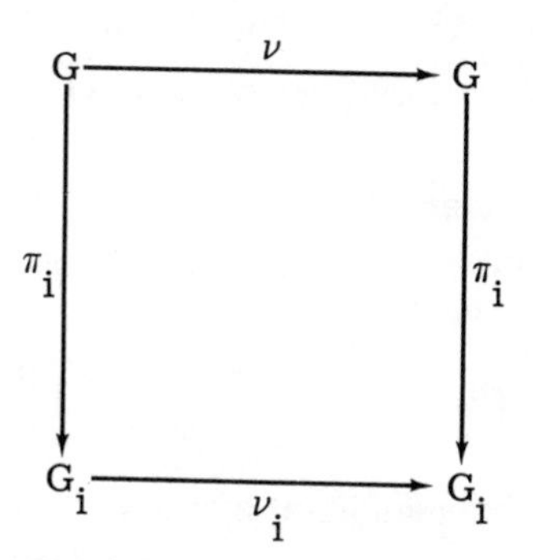

whence, as π_i, μ_i, ν_i are continuous for all i, so are $\mu\pi_i$ and $\nu\pi_i$. Thus μ, ν are continuous and G <u>is a topological group.</u>

Now given morphisms $f_i : H \to G_i$ in $\mathcal{TG}$, the unique induced map of sets

$$f : H \to G \text{ with } f\pi_i = f_i \text{ for each } i$$

is a group-morphism (product in $\mathcal{G}$) and continuous (product in $\mathcal{T}$), hence a $\mathcal{TG}$-morphism. It follows that G is the product of the G_i in $\mathcal{TG}$.

Examples. (i) Let $\mathbf{R}$ be the additive group of real numbers with the usual topology. Then

$$\mathbf{R}^n \cong \mathbf{R} \times \ldots \times \mathbf{R} \quad (n \text{ copies})$$

in $\mathcal{TG}$, that is, the product topology coincides with the metric topology on real n-space. The reason for this is that every open ball in $\mathbf{R}^n$ contains an open n-cube centred on the same point, and vice versa.

(ii) S^1 is an Abelian topological group, whence so is $T^n = S^1 \times \ldots \times S^1$ (n copies). (By definition T^0 is the trivial topological group with one element.) We observe that $S^1 \cong \mathbf{R}/\mathbf{Z}$ in $\mathcal{TG}$ or, what is the same thing, the morphism $\phi : \mathbf{R} \to S^1$ with kernel $\mathbf{Z}$ defined by $t \mapsto \exp 2\pi it$ is open. The product morphism $\phi^n : \mathbf{R}^n \to T^n$ is therefore open, and its kernel is $\mathbf{Z}^n$. It follows, by Proposition 1 ($\mathcal{TG}$) that $T^n \cong \mathbf{R}^n/\mathbf{Z}^n$ in $\mathcal{TG}$. More generally, if $G = \Pi G_i$ in $\mathcal{TG}$ and H_i is a normal subgroup of G_i for each i, then $H = \Pi H_i$ is a normal topological subgroup of G and $G/H \cong \Pi(G_i/H_i)$ in $\mathcal{TG}$. One simply needs the facts that a product of subspaces is a subspace of the product of the containing spaces, and that any product of open maps is open.

Exercises. (i) If $G = A \times B$ in $\mathcal{TG}$, let $A_0 = \{(a, e_B) \in G;\ a \in A\}$. Prove that A_0 is a normal subgroup of G and that $A_0 \cong A$, $G/A_0 \cong B$ in $\mathcal{TG}$.

(ii) Prove that $\mathbf{C}^* \cong \mathbf{R}^{pos} \times S^1$ in $\mathcal{TG}$.

4. Fundamental systems of neighbourhoods

Definition. Let X be a topological space and let $A \subseteq X$. An element $a \in A$ is said to be an interior point of A if there is an open set N of X with $a \in N \subseteq A$.

Definition. A neighbourhood of $x \in X$ is a subset of X containing x as an interior point. (An open neighbourhood of x is then any open set containing x.)

Definition. A fundamental system of neighbourhoods of x in X (f. s. n. of x) is a collection $\mathfrak{F}$ of neighbourhoods of x such that every neighbourhood of x contains a member of $\mathfrak{F}$. If each member of $\mathfrak{F}$ is open, we speak of a fundamental system of open neighbourhoods of x, (f. s. o. n. of x). (E. g. in $\mathbf{R}$ the intervals $(-1/n, 1/n)$ for $n \in \mathbf{N}$ form a f. s. o. n. of 0.)

Clearly, if for each $x \in X$ we are given a fundamental system of open neighbourhoods of x, $\mathfrak{F}(x)$ say, then $\bigcup_{x \in X} \mathfrak{F}(x)$ is a basis for the open sets of X.

Now suppose that G is a topological group and $\mathfrak{F}$ is a fundamental system of open neighbourhoods of the identity element e. Then for $x \in G$, $\mathfrak{F}(x) = \{Ux;\ U \in \mathfrak{F}\}$ is a fundamental system of open neighbourhoods of x (because r_x is a homeomorphism!).

Any fundamental system $\mathfrak{F}$ of open neighbourhoods of e in G has the following properties:

FN1: If $U, V \in \mathfrak{F}$, then $\exists W \in \mathfrak{F}$ such that $W \subseteq U \cap V$.

FN2: If $a \in U \in \mathfrak{F}$ then $\exists V \in \mathfrak{F}$ such that $Va \subseteq U$.

FN3: If $U \in \mathfrak{F}$, then $\exists V \in \mathfrak{F}$ such that $V^{-1}V \subseteq U$.

FN4: If $U \in \mathfrak{F}$, $x \in G$, then $\exists V \in \mathfrak{F}$ such that $x^{-1}Vx \subseteq U$.

Proof. In (1) $U \cap V$, in (2) Ua^{-1}, are open neighbourhoods of e, so contain an element of $\mathfrak{F}$.

(3): The map $f : G \times G \to G$ given by $(a, b) \mapsto a^{-1}b$ is continuous. Thus Uf^{-1} is open, contains (e, e), and hence contains a set of the form $A \times B$, where A, B are open and contain e. Since $A \cap B$ is an open neighbourhood of e, $\exists V \in \mathfrak{F}$ so that $V \subseteq A \cap B$. For this V we have $V \times V \subseteq Uf^{-1}$, that is, $V^{-1}V \subseteq U$.

(4): The map $f : G \to G$ given by $a \mapsto x^{-1}ax$ is continuous, so Uf^{-1} is open and contains e, hence contains some $V \in \mathfrak{F}$. For this V we have $U \supseteq Vf = x^{-1}Vx$.

Notes. (i) Any fundamental system of open neighbourhoods of e, say $\mathfrak{F}$, also satisfies: $\forall U \in \mathfrak{F}$, $\exists V, W \in \mathfrak{F}$ such that $V^{-1} \subseteq U$, $W^2 \subseteq U$. (We shall write W^2 for WW when there is no risk of confusion with $W \times W$.)

(ii) The set $\mathfrak{F}$ of all open neighbourhoods of e is a fundamental system of open neighbourhoods of e, and has the extra properties that if $U, V \in \mathfrak{F}$ then $U \cap V \in \mathfrak{F}$ and $U^{-1} \in \mathfrak{F}$. These imply that every $U \in \mathfrak{F}$ contains a symmetric $W \in \mathfrak{F}$ namely $W = U \cap U^{-1}$. (A neighbourhood W of e is called symmetric if $W^{-1} = W$.)

Proposition 2. *Let* G *be an abstract group and let* $\mathfrak{F}$ *be a non-empty collection of subsets of* G, *each containing* e, *such that* $\mathfrak{F}$ *satisfies the conditions* **FN1-FN4**. *Then there is a unique topology on* G *such that* G *is a topological group and* $\mathfrak{F}$ *is a fundamental system of open neighbourhoods of* e.

Proof. Uniqueness is clear, because if $\mathfrak{F}$ is a fundamental system of open neighbourhoods of e, then $\mathcal{B} = \{Ug;\ U \in \mathfrak{F},\ g \in G\}$ is a basis for the topology. So, for given $\mathfrak{F}$, we first check the basis conditions for $\mathcal{B}$ (see p. 13).

B1: $\mathfrak{F}$ is non-empty, so $\exists U \in \mathfrak{F}$, $e \in U$. Thus $g \in Ug$, and $\bigcup_{g \in G} Ug = G$.

B2: Let $U, V \in \mathfrak{F}$, $a, b \in G$, and let $c \in Ua \cap Vb$. (If $Ua \cap Vb = \phi$, we are there.) Let $u = ca^{-1} \in U$, $v = cb^{-1} \in V$. Then (by **FN2**) $\exists U_1, V_1 \in \mathfrak{F}$ such that $U_1 u \subseteq U$, $V_1 v \subseteq V$, whence $U_1 c \subseteq Ua$, $V_1 c \subseteq Vb$. By **FN4** we can choose $W \in \mathfrak{F}$, $W \subseteq U_1 \cap V_1$. Then $Wc \subseteq Ua \cap Vb$. Hence the sets $\{Ug : U \in \mathfrak{F},\ g \in G\}$ do give a basis for the open sets of a topology on G.

To show that G is a topological group under this topology, it is enough to show that the map $f : G \times G \to G$, defined by $(a, b) \mapsto a^{-1}b$ is continuous. Since $\mathcal{B}$ is an open basis, we need only show that $(Ua)f^{-1}$ is open, for $U \in \mathfrak{F}$, $a \in G$.

Let $(b, c) \in (Ua)f^{-1}$, i.e., $b^{-1}c = ua$ for some $u \in U$. By **FN2**, $\exists V \in \mathfrak{F}$ such that $Vu \subseteq U$, and by **FN4**, $\exists W \in \mathfrak{F}$ such that $b^{-1}Wb \subseteq V$. For this W we have $b^{-1}Wca^{-1} \subseteq U$. By **FN3**, $\exists Z \in \mathfrak{F}$ such that $Z^{-1}Z \subseteq W$, and then $b^{-1}Z^{-1}Zca^{-1} \subseteq U$. Thus $(Zb)^{-1}(Zc) \subseteq Ua$, i.e. $Zb \times Zc \subseteq (Ua)f^{-1}$, which is therefore open.

Corollary. *Any group* G *can be made into a topological group by specifying, as fundamental system of open neighbourhoods of* e, *a set* $\mathfrak{F}$ *of subgroups satisfying:*

FS1: *If* $U, V \in \mathfrak{F}$, *then* $\exists W \in \mathfrak{F}$ *such that* $W \subseteq U \cap V$.

FS2: *If* $U \in \mathfrak{F}$, $x \in G$, *then* $\exists V \in \mathfrak{F}$ *such that* $x^{-1}Vx \subseteq U$.

(**Proof.** **FN2**, **FN3** hold automatically with $V = U$.)

Thus any collection of subgroups of G containing all conjugates of its members and all finite intersections of its members (in particular any chain of normal subgroups of G), defines a topological group structure on G, in which the open sets are all unions of right (or left) cosets of the given subgroups.

Examples. (i) **Q** (the additive group of rationals) is a topological group under the p-adic topology. If, for $t \in \mathbf{Z}$, we define $U_t = \{mp^t/n; p \nmid n\}$, then we have a countable chain of (normal) subgroups

$$\ldots \supseteq U_{-1} \supseteq U_0 \supseteq U_1 \ldots$$

giving a fundamental system of open neighbourhoods of 0 for this topology.

(ii) In $\mathbf{Q}^*$ under the p-adic topology, the subsets $V_t = 1 + U_t$, for $t \in \mathbf{Z}$ form a fundamental system of open neighbourhoods of 1.

(iii) For any abstract group G, let $\mathfrak{F}$ be the set of all subgroups of finite index. Then $\mathfrak{F}$ determines a topological group structure on G such that $\mathfrak{F}$ is a fundamental system of open neighbourhoods of e.

(iv) In any group, the terms of the lower central series (or the derived series) similarly determine a topological group structure.

(v) If R is any commutative ring, let $R\langle x\rangle$ be the power-series ring, with elements all formal (infinite) sums $\sum_{n=0}^{\infty} a_n x^n$, where $a_n \in R$. Define additive subgroups $U_n = x^n R\langle x\rangle$, for all n. Then $U_0 \supseteq U_1 \supseteq U_2 \supseteq \ldots$, and this sequence defines a topology which makes $R\langle x\rangle$ an (additive) topological group. (Exercise: $R\langle x\rangle$ is in fact a topological ring, i.e., multiplication of power series is continuous with the topology thus defined.)

5. Separation axioms

Definition. Let X be a topological space.

(i) X is a T_1-space if for any pair of distinct points a, b in X there exists a neighbourhood of a not containing b.

(ii) X is a T_2-space (or a Hausdorff space) if for any pair of distinct points a, b in X there exist neighbourhoods of a and b which are disjoint.

Clearly X is a T_1-space if and only if every one-point subset is closed.

Note that every discrete space is Hausdorff, and every Hausdorff space is a T_1-space.

Proposition 3 ($\mathcal{T}$). If X is a topological space then the following are equivalent:

(i) X is Hausdorff.

(ii) The diagonal map $\delta : X \to X \times X$ given by $x \mapsto (x, x)$ is a closed map.

(iii) For any pair of continuous maps $f, g : Y \to X$, their difference kernel $Z = \{y; yf = yg\}$ is a closed subset of Y.

Proof. (i) $\Rightarrow$ (ii): Let Y be a closed subset of X and let $(a, b) \notin Y\delta$. Either (1) $a \neq b$ or (2) $a = b$ and $a \notin Y$.

In case (1) there exist disjoint open sets A and B containing a and b respectively and then $A \times B$ is a neighbourhood of (a, b) in $X \times X$ which does not meet $Y\delta$.

In case (2), since Y is closed, there exists an open set A containing a which is disjoint from Y. In this case $A \times A$ is an open neighbourhood of $(a, b) = (a, a)$ in $X \times X$ not meeting $Y\delta$. Hence $Y\delta$ is closed and δ is a closed map.

(ii) $\Rightarrow$ (iii): Let $f, g : Y \to X$ be in $\mathcal{T}$. Since X is closed and δ is a closed map, $\Delta = X\delta$ is a closed subset of $X \times X$. Now f and g induce a continuous map $(f, g) : Y \to X \times X$ and the difference kernel Z of f and g is the pre-image under (f, g) of the closed set Δ, so Z is closed.

(iii) $\Rightarrow$ (i): Taking f and g to be the projections $X \times X \to X$, their difference kernel Δ is closed. Hence if $a \neq b$ are points of X, there exists an open neighbourhood N of (a, b) in $X \times X$ which is disjoint from Δ. We may now choose open neighbourhoods A and B of a and b respectively in X so that $A \times B$ is contained in N and so is disjoint from Δ, giving $A \cap B = \emptyset$ as required.

Proposition 3 ($\mathcal{TG}$). Let G be a topological group and $\mathfrak{F}$ a fundamental system of neighbourhoods of e in G. Then the following statements are equivalent:

(i) G <u>is Hausdorff.</u>

(ii) $\delta : G \to G \times G$ <u>is a closed map.</u>

(iii) <u>For</u> $f, g : H \to G$ <u>in</u> $\mathcal{T}\mathcal{G}$, <u>their difference kernel is a closed subgroup of</u> H.

(iv) <u>For any</u> $f : H \to G$ <u>in</u> $\mathcal{T}\mathcal{G}$, Ker f <u>is a closed subgroup of</u> H.

(v) $\{e\}$ <u>is a closed subgroup of</u> G.

(vi) G <u>is</u> T_1, <u>i. e., every one-point subset of</u> G <u>is closed.</u>

(vii) $\cap \mathcal{F} = \{e\}$.

(viii) <u>The intersection of all neighbourhoods of</u> e <u>is</u> $\{e\}$.

Proof. (i) $\Rightarrow$ (ii) follows directly from Proposition 3 ($\mathcal{T}$) whilst (ii) $\Rightarrow$ (iii) requires only the extra result that the difference kernel of two group homomorphisms is a subgroup.

(iii) $\Rightarrow$ (iv): Ker f is the difference kernel of f and g, where $g : H \to G$ is defined by $h \mapsto e$ for all h in H.

(iv) $\Rightarrow$ (v): Take f as the identity map on G.

(v) $\Rightarrow$ (vi): G is a homogeneous space.

(vi) $\Rightarrow$ (vii): Let $g \neq e$ be an element of G. Since $\{g\}$ is closed and $e \notin \{g\}$ there exists $U \in \mathcal{F}$ such that $U \cap \{g\} = \emptyset$. Hence $g \notin \cap \mathcal{F}$ and so $\cap \mathcal{F} = \{e\}$.

(vii) $\Rightarrow$ (viii) is trivial.

(viii) $\Rightarrow$ (i): Let $a \neq b$ be elements of G. Since $ab^{-1} \neq e$ there exists a neighbourhood U of e which does not contain ab^{-1}. We may now find a neighbourhood V of e such that $V^{-1}V \subseteq U$. Then $ab^{-1} \notin V^{-1}V$, so $Vab^{-1} \cap V = \emptyset$ and $Va \cap Vb = \emptyset$. But Va and Vb are neighbourhoods of a and b respectively and so G is Hausdorff.

Definition. A space X is a T_0-space if, given any two distinct points a and b in X, there exists an open set containing exactly one of them.

Exercise. Every T_0 topological group is Hausdorff.

Another property of spaces akin to the separation properties T_0, T_1, T_2 is regularity. A space X is <u>regular</u> if every neighbourhood of each point x contains a closed neighbourhood of x. This is easily seen to be equivalent to the following condition: for every closed set A in X and every point $b \notin A$, there exist open sets U, V in X such that $A \subseteq U$ $b \in V$ and $U \cap V = \emptyset$. Thus a regular T_1-space is Hausdorff since we may take A to be a point.

Exercise. Prove that every regular T_0-space is Hausdorff.

In view of the fact that topological groups need not be T_0-spaces it is somewhat surprising to find:

Proposition 4. (i) Every topological group G is regular.

(ii) Every left coset space G/H is regular.

Proof. (i) By homogeneity it is enough to show that if A is closed and $e \notin A$ then there exist open sets U, V with $A \subseteq U$, $e \in V$ and $U \cap V = \emptyset$. Now the complement of A is a neighbourhood of e; we can therefore find (by **FN3**) an open neighbourhood V of e such that $V^{-1}V \cap A = \emptyset$. But this implies $V \cap VA = \emptyset$, so we may take $U = VA$ which contains A and is open by Proposition 0(iv).

(ii) The proof is similar. Given a closed set A in G/H and a point xH not in it, let $\tilde{A} = Aq^{-1}$, where q is the quotient map $G \to G/H$. Then $\tilde{A}$ is closed in G, and $x \notin \tilde{A}$; we can therefore find an open neighbourhood V of e in G such that $V^{-1}Vx \cap \tilde{A} = \emptyset$, and hence $Vx \cap V\tilde{A} = \emptyset$. Now $\tilde{A} = Aq^{-1}$ is a union of left cosets of H; therefore so is $V\tilde{A}$, and we deduce that in fact $VxH \cap V\tilde{A} = \emptyset$. It follows that the images in G/H of the open sets VxH and $V\tilde{A}$ are disjoint open sets containing xH and A, respectively.

Proposition 5 ($\mathcal{T}$). (i) If X is a Hausdorff space and $f : Y \to X$ is a continuous injection then Y is Hausdorff. In particular, subspaces of Hausdorff spaces are Hausdorff.

(ii) If $\{X_i\}_{i \in I}$ is a family of non-empty spaces, then $X = \Pi X_i$ is Hausdorff if and only if all the X_i are Hausdorff.

Proof. (i) If $a \neq b$ are points of Y then $af \neq bf$. Let A and B be disjoint neighbourhoods of af and bf respectively; then their pre-images are disjoint neighbourhoods of a and b.

(ii) Suppose that X is Hausdorff and let $j \in I$. For each $i \in I$ define $f_i : X_j \to X_i$ in $\mathcal{T}$ as follows: (a) f_j is the identity map on X_j;

(b) for $i \neq j$ pick some fixed element $x_i \in X_i$ and let f_i be the corresponding constant map. These f_i induce an injection $f : X_j \to X$ in $\mathcal{T}$ and so by part (i) X_j is Hausdorff.

Conversely, suppose that each X_i is Hausdorff and that a and b are distinct points in X, with coordinates $\{a_i\}$ and $\{b_i\}$ respectively. Since $a \neq b$ there exists an i such that $a_i \neq b_i$ and hence there exist disjoint neighbourhoods A and B of a_i and b_i respectively in X_i. Clearly $A\pi_i^{-1}$ and $B\pi_i^{-1}$ are disjoint neighbourhoods of a and b in X.

Proposition 5 ($\mathcal{T}\mathcal{G}$). Let G, G_i be topological groups and let H be a subgroup of G. Then:

(i) G Hausdorff $\Rightarrow$ H Hausdorff;

(ii) G/H is Hausdorff $\iff$ H is a closed subgroup;

(iii) H Hausdorff and G/H Hausdorff $\Rightarrow$ G Hausdorff;

(iv) ΠG_i is Hausdorff $\iff$ every G_i is Hausdorff.

Proof. (i) and (iv) follow from Proposition 5 ($\mathcal{T}$).

(ii): By the definition of quotient topology, G/H is a T_1-space if and only if H (and therefore all its cosets) are closed in G. If H is a normal subgroup then G/H is a group and the result follows from Proposition 3 ($\mathcal{T}\mathcal{G}$). If H is not normal we need a slight elaboration of the argument of Proposition 3 ($\mathcal{T}\mathcal{G}$), as follows.

If G/H is Hausdorff then it is T_1, so H is closed. Conversely, if H is closed, and aH, bH are distinct cosets, it suffices to find a neighbourhood W of e in G such that $WaH \cap WbH = \emptyset$. Since $a^{-1}bH$ is closed and does not contain e, we can find neighbourhoods U, V, W of e in G such that $U \cap a^{-1}bH = \emptyset$, $V^{-1}V \subseteq U$ and $a^{-1}Wa \subseteq V$. It is easy to check that W has the required property.

(iii): If G/H is Hausdorff, H is closed in G by (ii). If also H is Hausdorff then $\{e\}$ is closed in H. Hence $\{e\}$ is closed in G, and G is Hausdorff.

Examples. (i) **R** and all its subgroups are Hausdorff.

(ii) Since **Z** is closed in **R**, $S^1 \cong \mathbf{R}/\mathbf{Z}$ is Hausdorff. (Alternatively, S^1 is Hausdorff because it is a subgroup of the Hausdorff group $\mathbf{C}^*$.)

(iii) $T^n = S^1 \times S^1 \times \ldots \times S^1$ is a Hausdorff group.

(iv) $\mathbf{R}/\mathbf{Q}$ is not a Hausdorff group since $\mathbf{Q}$ is not closed in $\mathbf{R}$.

(v) If G is any topological group, let E be the closure of the trivial subgroup, that is, the intersection of all closed sets containing e. Then E is closed, and we shall show that it is a normal subgroup of G. The sets E^{-1}, $x^{-1}E$ for $x \in E$, and $y^{-1}Ey$ for $y \in G$ are all closed in G, being images of E under homeomorphisms $G \to G$; they also all contain e. Hence $E \subset E^{-1}$, $E \subset x^{-1}E$ for $x \in E$, and $E \subset y^{-1}Ey$ for $y \in G$. It follows easily that $E^{-1} \subset E$, $xE \subset E$ for $x \in E$, and $yEy^{-1} \subset E$ for $y \in G$, whence E is a normal subgroup. The group G/E is a Hausdorff group since E is closed.

Exercises. (i) Show that G/E is the universal Hausdorff group on G, i.e., for any morphism $f : G \to H$ in $\mathcal{TG}$, where H is Hausdorff, $\exists!\ f^* : G/E \to H$ in $\mathcal{TG}$ such that $f = qf^*$, where $q : G \to G/E$ is the quotient map.

(ii) Prove also that if G^0 denotes the group G with its topology replaced by the trivial topology then the map $g \mapsto (gE, g)$ from G to $(G/E) \times G^0$ embeds G as a topological subgroup in $(G/E) \times G^0$.

6. Open subgroups

Proposition 6. Let G be a topological group. Then:

(i) every open subgroup of G is closed and every closed subgroup of finite index is open;

(ii) every subgroup of G containing a neighbourhood of e is open;

(iii) if H is a subgroup of G then G/H is discrete if and only if H is open.

Proof. (i) If H is open in G then by Proposition 0 so are all cosets of H. Also the complement of H in G is a union of cosets and hence is open. Therefore H is closed. Similarly if H is closed of finite index, its complement is the union of a finite number of cosets, each closed. Hence H is open.

(ii) If H contains a neighbourhood of e it contains an open neighbourhood U or e. Hence $H = HU$, which is open by Proposition 0.

(iii) G/H is discrete $\iff$ all points are open in G/H

$\iff$ H and all left cosets of H are open in G

$\iff$ H is open in G.

7. Connectedness

Definition. A topological space X is <u>connected</u> if it is non-empty and cannot be written $X = A \cup B$ where A and B are non-empty, disjoint open subsets of X. Thus a non-empty space X is connected if and only if the only subsets of X which are both open and closed are $\emptyset$ and X.

A <u>disconnection</u> of X is a pair of complementary, non-empty, open-closed (i.e. open <u>and</u> closed) subspaces.

Corollary to Proposition 6. (i) <u>A connected topological group has no proper open subgroups.</u>

(ii) <u>A connected topological group is generated as an abstract group by any neighbourhood of</u> e <u>(or by any non-empty open subset)</u>.

Proposition 7 ($\mathcal{T}$). (i) <u>If</u> X <u>is connected and</u> $f : X \to Y$ <u>is a continuous function then</u> Xf <u>is connected.</u>

(ii) ΠX_i <u>is connected if and only if every</u> X_i <u>is connected.</u>

Proof. (i) Let $Z = Xf$ and suppose that A is an open-closed subset of Z then Af^{-1} is open-closed in X. Hence Af^{-1} is $\emptyset$ or X and so A is $\emptyset$ or Z.

(ii) If $X = \Pi X_i$ is connected then so are the X_i by (i). Suppose now that all the X_i are connected and that A is a non-empty open-closed subset of X. We show that $A = X$.

Let $a = \{a_i\} \in A$ and let $j \in I$. We define $\theta : X_j \to X$ to be the map whose components are the constant maps $\theta_i : X_j \to X_i$ ($x_j \mapsto a_i$) for $i \neq j$ and the identity map on X_j. The image $X_j\theta$ is connected by (i) and contains a. Hence $A \cap X_j\theta$ is non-empty and open-closed in $X_j\theta$, so $A \supseteq X_j\theta$. Therefore, if a' differs from a in one component then $a' \in A$. Clearly, by induction, if a' differs from a in a finite number of components and $a \in A$, then $a' \in A$. Now A is open in X so there exists a basic neighbourhood of a inside A. Suppose this is ΠU_i where

$a_i \in U_i$, U_i is open in X_i and for all but a finite number of i, $U_i = X_i$. Given any $x \in X$, by changing a finite number of components of x we obtain a point in ΠU_i and hence in A. Hence $x \in A$ and it follows that $A = X$.

Proposition 7 ($\mathcal{T}\mathcal{G}$). Let G, G_i be topological groups and let H be a subgroup of G. Then

(i) G connected $\Rightarrow$ G/H connected.

(ii) H connected and G/H connected $\Rightarrow$ G connected.

(iii) ΠG_i is connected $\Longleftrightarrow$ every G_i is connected.

Proof. (i) and (iii) follow from Proposition 7 ($\mathcal{T}$).

(ii) Suppose that H and G/H are connected and that G has a disconnection $G = A \cup B$, where we may assume that $e \in A$. Since H is connected so are all its left cosets. Now each coset meets either A or B and so must be contained in one of them (because A, B are open-closed). Therefore A and B are each a union of left cosets of H. If $q : G \to G/H$ is the quotient map, it follows that $Aq \cap Bq = \emptyset$. Since q is an open map, Aq and Bq are open-closed, so we have constructed a disconnection $Aq \cup Bq$ of G/H. This is a contradiction, so G is connected.

Note: A subgroup of a connected group need not be connected. E.g., **R** is connected but **Z** and **Q** are not.

Proposition 8 ($\mathcal{T}$). If X is a topological space and $x \in X$ then the union of all connected subspaces of X containing x is connected and is closed in X.

(This subspace will be called the component of X at x and will be denoted by $\mathrm{Comp}_X x$ or $\mathrm{Comp}\, x$.)

Proof. Let C_i, $i \in I$, be all the connected subspaces of X which contain x, and let $C = \bigcup_{i \in I} C_i$. Then $C \neq \emptyset$ since $\{x\}$ is connected.

Suppose that C has a disconnection $A \cup B$ where $x \in A$. Since $A \cap C_i \neq \emptyset$ and C_i is connected we have $A \supseteq C_i$ for all i, which implies that $A \supseteq C$, a contradiction. Hence C is connected.

Now let $\overline{C}$ be the closure of C in X, i.e., the intersection of all closed subspaces of X containing C. We shall verify that $\overline{C}$ is connected, which will imply that $\overline{C} \subseteq C$ thus proving that C is closed.

Suppose that $\overline{C} = A \cup B$ is a disconnection of $\overline{C}$ with $x \in A$. Then $A \cap C \neq \emptyset$ and C is connected, so $A \supseteq C$. Now A is closed in $\overline{C}$ which is closed in X, so A is closed in X. Since $A \supseteq C$, this implies that $A \supseteq \overline{C}$, so $A \cup B$ is not a disconnection of $\overline{C}$. Hence $\overline{C}$ is connected.

Corollary. The components of X form a partition of X.

Proof. Certainly $\bigcup_{x \in X} \text{Comp } x = X$ since $x \in \text{Comp } x$.

Suppose that $\text{Comp } x \cap \text{Comp } y \neq \emptyset$, say $z \in \text{Comp } x \cap \text{Comp } y$. Since Comp x is connected and contains z we have $\text{Comp } x \subseteq \text{Comp } z$. But this implies that x is contained in Comp z, so $\text{Comp } z \subseteq \text{Comp } x$. Hence $\text{Comp } x = \text{Comp } z = \text{Comp } y$.

N. B. Components are not in general open, so a single component and its complement do not in general give a disconnection of a space. If a space has only a finite number of components then these are all open and closed. In this case every division of the components into two classes gives a disconnection of the space.

Definition. A space X is said to be totally disconnected if each component of X has just one point, i. e., there do not exist connected subspaces containing more than one point. E. g., a discrete space is totally disconnected. (But note: totally disconnected does not imply discrete!)

Proposition 9 ($\mathcal{T}$). Let X, X_i be topological spaces, with $X_i \neq \emptyset$ for each i. Then

(i) if X is totally disconnected then every subspace of X is totally disconnected;

(ii) ΠX_i is totally disconnected if and only if each X_i is totally disconnected.

Proof. (i) is obvious.

(ii) Suppose that ΠX_i is totally disconnected and $X_i \neq \emptyset$ for each i. Let A_i be a connected subspace of X_i for each i. Then ΠA_i is connected (by Proposition 7 $(\mathcal{T})$) and is a subspace of ΠX_i (see Exercise (iv), p. 15), thus can only have one point. Therefore each A_i has only one point, i.e., each X_i is totally disconnected.

Conversely suppose that $X_i \neq \emptyset$, and X_i is totally disconnected for each i. Let $A \subseteq \Pi X_i$ be connected. Then $A\pi_i$ is a connected subspace of X_i (by Proposition 7 $(\mathcal{T})$), so has only one point. Therefore A has only one point, i.e., ΠX_i is totally disconnected.

Corollary. *Any product of discrete spaces is totally disconnected.*

Proposition 9 $(\mathcal{T}\mathcal{G})$. *Let* G, G_i *be topological groups, and* H *a subgroup of* G. *Then*

(i) *if* G *is totally disconnected then so is* H;

(ii) *if* G/H *and* H *are totally disconnected then* G *is totally disconnected;*

(iii) ΠG_i *is totally disconnected if and only if each* G_i *is totally disconnected.*

Proof. (i) and (iii) are contained in Proposition 9 $(\mathcal{T})$.

(ii) Let A be a connected subspace of G. Let $q : G \to G/H$ be the quotient map. Then Aq is connected, so has just one element, that is, A is contained in some left coset of H. But each coset of H, being homeomorphic to H, is totally disconnected. Hence A has just one element.

Theorem 1. *Let* G *be a topological group and let* H *be the component of* G *at* e. *Then* H *is a closed, connected, normal subgroup of* G, *and* G/H *is a totally disconnected (Hausdorff) group. The components of* G *are just the cosets of* H. *Every open subgroup of* G *contains* H.

Proof. By Proposition 8, H is connected and closed in G. For any $h \in H$ and any $g \in G$, the sets $h^{-1}H$ and $g^{-1}Hg$ are connected, being homeomorphic to H; also $e \in h^{-1}H$ and $e \in g^{-1}Hg$. Since H is the component of G at e, this implies that $h^{-1}H \subseteq H$ and $g^{-1}Hg \subseteq H$ for all $h \in H$ and all $g \in G$, which shows that H is a normal subgroup of G. Now G/H is a topological group, Hausdorff since H is closed.

Consider its component at its identity element. This is a normal subgroup of G/H and so, by the corollary to Proposition 1, it is isomorphic in $\mathcal{TG}$ to a group L/H, where L is a topological subgroup of G containing H. Now L/H and H are both connected, so L is connected, by Proposition 7 ($\mathcal{TG}$). It follows that $L \subseteq H$ and so the component of G/H at its identity element is trivial.

The component of G at an arbitrary point x is just xH, because l_x is a homeomorphism from G to G. Similarly, in G/H, the components are just the cosets of the trivial subgroup, i.e., one-point sets. Hence G/H is totally disconnected.

Finally, if U is an open subgroup of G, then U is open and closed (Proposition 6) and contains e. Hence $U \cap H$ is a non-empty, open-closed subset of the connected space H, and so must be the whole of H, i.e., $U \supseteq H$.

Note: Any totally disconnected group is Hausdorff because the trivial subgroup is a component and is therefore closed (see Propositions 8 ($\mathcal{T}$) and 3 ($\mathcal{TG}$)).

Examples. (i) $(\mathbf{R}, +)$ is a connected group. For suppose that X is a non-empty open-closed subset of $\mathbf{R}$, with $X \neq \mathbf{R}$. Let $x \in X$ and let A be the set of all real numbers $a \geq 0$ such that X contains the closed interval $[x-a, x+a]$. Then $0 \in A$ and A is bounded above (otherwise $X = \mathbf{R}$). Hence A has a least upper bound m. If $m \in A$ then $[x-m, x+m] \subseteq X$ and since X is open, X contains a neighbourhood of each of $x - m$, $x + m$ and so contains an interval $[x-m', x+m']$ with $m' > m$, which is impossible. Hence $m \notin A$ and therefore one of $x - m$, $x + m$ is not in X, say $x + m \notin X$. Since X is closed, some neighbourhood of $x + m$ does not meet X, so $\exists m'$, $0 \leq m' < m$, such that $[x-m', x+m'] \subsetneq X$. This m' is an upper bound for A, and we have the required contradiction since m is the least upper bound.

(ii) $(\mathbf{R}^*, \cdot)$ has two components, because $\mathbf{R}^{pos}$, being isomorphic in $\mathcal{TG}$ to $(\mathbf{R}, +)$, is connected, and is an open subgroup. The quotient $\mathbf{R}^*/\mathbf{R}^{pos}$ is a discrete group of order two.

(iii) $(\mathbf{R}^n, +)$ is a connected group by Example (i) and Proposition 7 ($\mathcal{TG}$). Alternatively we may argue as follows. A space X is called path-connected if, given $a, b \in X$, there is a continuous function $p : I \to X$ (where I denotes the closed unit interval $[0, 1]$ in $\mathbf{R}$) such that $0p = a$, $1p = b$. Any path-connected space is connected. For if $X = A \cup B$ is a disconnection of X, we may choose $a \in A$, $b \in B$ and obtain a path from a to b as described above. Then $I = Ap^{-1} \cup Bp^{-1}$ is a disconnection of I. This is impossible because I is connected (Exercise). Now in $\mathbf{R}^n$, any two points a and b can be joined by a line-segment, which is a continuous image of I under the map $p : t \mapsto (1 - t)a + tb$. Hence $\mathbf{R}^n$ is path-connected.

(iv) $\mathbf{R}^n \setminus \{0\}$ is connected if $n \geq 2$. Indeed, it is path-connected because any two points can be joined to a third point by line segments not passing through 0.

(v) S^n is connected if $n \geq 1$ since there is a continuous surjection $x \mapsto x/|x|$ from $\mathbf{R}^{n+1} \setminus \{0\}$ to S^n. (Note: S^0 is a discrete two-point space.)

(vi) T^n is connected for all $n \geq 0$ since there is a continuous surjection from $\mathbf{R}^n$ to T^n. Alternatively, it is connected because it is a product of connected groups S^1.

(vii) In a free group F, the lower central series $F = F_1 \supset F_2 \supset \ldots$ is a chain of normal subgroups. (F_n is generated by all commutators of weight n.) By the corollary to Proposition 2, F is a topological group with respect to a topology in which $\{F_n\}$ is a fundamental system of open neighbourhoods of e. Now Magnus has proved that $\cap F_n = \{e\}$ (see 'Combinatorial Group Theory' by Magnus, Karrass and Solitar (Interscience), Theorem 5.7 and Corollary). Since the groups F_n are open in F, they all contain the component of F at e (Theorem 1) and it follows that F is totally disconnected.

(viii) $\mathbf{Q}$, as a subgroup of $\mathbf{R}$ is totally disconnected. For let $a, b \in \mathbf{Q}$ with $a < b$ and suppose that X is a connected subspace of $\mathbf{Q}$ containing a and b. Then $\exists\, r \in \mathbf{R}$ such that $a < r < b$ and $r \notin \mathbf{Q}$. The sets $A = \{x \in X;\ x < r\}$ and $B = \{x \in X;\ x > r\}$ are now easily seen to form a disconnection of X.

(ix) $\mathbf{Q}$, with the p-adic topology, has as a fundamental system of open neighbourhoods of 0 the chain of subgroups

$$U \supset pU \supset p^2U \supset \ldots \supset p^rU \supset \ldots, \text{ where } U = \{m/n;\ p \nmid n\}.$$

The intersection of these subgroups is $\{0\}$, so $\mathbf{Q}$ is totally disconnected as in Example (vii).

(x) The orthogonal group O_n is the group of all $n \times n$ real matrices M such that $M^t M = I_n$. These matrices represent linear transformations $\mathbf{R}^n \to \mathbf{R}^n$ which preserve distances (and therefore angles); the columns of such a matrix are orthogonal vectors of unit length. As a subspace of $\mathbf{R}^{n^2}$, O_n inherits a topology which makes it a topological group; it is a topological subgroup of $GL_n(\mathbf{R})$. Now the determinant map from O_n to $\mathbf{R}^*$ is continuous, since $\det M$ is a polynomial in the n^2 coordinates of M, and it is a group homomorphism. Its image is the group $\{\pm 1\}$ and its kernel is the group O_n^+ of all orthogonal matrices of determinant 1 (also denoted by SO_n in the literature). Since $\{\pm 1\}$ is a discrete subgroup of $\mathbf{R}^*$, O_n^+ is an open normal subgroup of O_n, so O_n is not connected. We shall prove that O_n^+ is connected and is therefore the component of O_n at the identity. Thus O_n has exactly two components.

To prove that O_n^+ is connected, we use induction on n. For $n = 1$ the assertion is trivially true. We assume, therefore, that O_n^+ is connected and consider the embedding $O_n^+ \to O_{n+1}^+$ given by

$$M \to \begin{pmatrix} M & O \\ O & 1 \end{pmatrix}$$

It is clear that O_n^+ is $\mathcal{TG}$-isomorphic to its image in O_{n+1}^+, so we may consider O_n^+ as a topological subgroup of O_{n+1}^+. If we can prove that the left coset space O_{n+1}^+/O_n^+ is connected, then Proposition 7 ($\mathcal{TG}$) and the induction hypothesis will imply that O_{n+1}^+ is connected, and the proof will be complete. In fact, we shall prove that O_{n+1}^+/O_n^+ is homeomorphic to the sphere S^n, which we have already shown to be connected.

For $M \in O_{n+1}^+$, let $\phi(M)$ denote the last column of M. This is a vector of length 1 in $\mathbf{R}^{n+1}$, so ϕ is a function from O_{n+1}^+ to S^n, and it is clearly continuous. Any vector of length 1 in $\mathbf{R}^{n+1}$ is part of an orthonormal basis (apply the Gram-Schmidt process to any basis starting with the given vector). It follows that any column vector of unit length is the last column of some orthogonal matrix, and hence the map $\phi : O_{n+1}^+ \to S^n$ is a surjection. If M_1 and M_2 are orthogonal matrices with the same last column, then $M_2 = M_1 M$ where M is of the form $\begin{pmatrix} A & O \\ u & 1 \end{pmatrix}$. Since $M \in O^+_{n+1}$ its last row has length 1, so $u = 0$ and $M \in O^+_n$. The converse is clear, and it follows that the fibres of $\phi : O^+_{n+1} \to S^n$ are precisely the left cosets of O^+_n. Thus ϕ induces a continuous bijection $\psi : O^+_{n+1}/O^+_n \to S^n$. Our proof will be complete if we can show that ψ is a homeomorphism and, by Proposition 1 ($\mathcal{T}$), it is therefore enough to prove that ϕ is a closed map. We postpone the proof of this final step until the next section so that we can use a compactness argument.

Exercises. (i) Prove that any product of path-connected spaces is path-connected.

(ii) Prove that $GL_n(\mathbf{C})$ is connected.

(iii) Prove that $GL_n(\mathbf{R})$ has two components.

(iv) Show that a quotient space of a totally disconnected space is not necessarily totally disconnected.

(v) (Harder) Show that a quotient group of a totally disconnected group is not necessarily totally disconnected.

(vi) Prove that if a finite topological group is connected then it must have the trivial topology. (Note: There exist finite connected spaces with non-trivial topology. An interesting example is the so-called 'small circle': let $X = \{a, b, c, d\}$ have open sets $\emptyset$, X, $\{a\}$, $\{c\}$, $\{a, c\}$, $\{a, b, c\}$ and $\{a, c, d\}$. Then X is clearly a topological space and is in fact a quotient space of S^1 under the map defined by: $1 \mapsto b$; $-1 \mapsto d$; $e^{2\pi it} \mapsto a$ for $0 < t < \frac{1}{2}$; $e^{2\pi it} \mapsto c$ for $\frac{1}{2} < t < 1$. It follows that X is path-connected. However, X has no group structure making it a topological group since it is not a homogeneous space.)

8. Compactness

Definition. A space X is compact if it has the Heine-Borel property: every cover of X by open sets, $X = \bigcup_{i \in I} X_i$, can be reduced to a finite sub-cover, i.e., there exist

$$i(1), i(2), \ldots, i(n) \in I \text{ such that } X = X_{i(1)} \cup X_{i(2)} \cup \ldots \cup X_{i(n)}.$$

Alternative definition (trivially equivalent to the above): X is compact if every family $\{C_i\}$ of closed subspaces of X has the 'finite intersection property', which asserts that if each finite subfamily of $\{C_i\}$ has non-empty intersection, then the whole family has non-empty intersection.

Notes: 1. In older books this property is called 'bicompactness'.

2. Many authors call a space 'compact' only if it is also Hausdorff.

3. X is said to be sequentially compact if (i) it has the Bolzano-Weierstrass property: every infinite subspace S of X has a point of accumulation (i.e., there exists $b \in X$ such that every neighbourhood of b contains an infinite subset of S), or, equivalently, (ii) every sequence of elements of X has a limit point (i.e., given $\{x_n\}$, $x_n \in X$, we can find $b \in X$ such that every neighbourhood of b contains x_n for infinitely many n). This property was called 'compactness' by earlier authors. [Exercises. (i) If X is compact then it is sequentially compact. (ii) If either (a) X is a metric space, or (b) X has a countable basis of open sets, then X is compact if and only if X is sequentially compact.]

4. A subspace of $\mathbf{R}^n$ is compact if and only if it is bounded and closed. This is the content of the Heine-Borel Theorem. In particular $[0, 1]$ is compact. More generally, a subspace of a complete metric space is compact if and only if it is closed and 'totally bounded' (i.e., for any given $\varepsilon > 0$, it can be covered by finitely many open ε-balls). For properties of metric spaces, see 'Functional Analysis' by Kolmogorov and Fomin (Graylock), Volume 1.

5. A discrete space is compact if and only if it is finite. (Exercise!)

Proposition 10 ($\mathcal{T}$). (i) Any closed subspace of a compact space is compact.

(ii) Any compact subspace of a Hausdorff space is closed.

(iii) The image of a compact space under a continuous map is compact.

(iv) The union of a finite number of compact subspaces of a space is compact.

(v) If C is compact, H is Hausdorff and $f : C \to H$ is continuous then f is a closed map.

(vi) _If_ $f : C \to H$ _as in (v) is a bijection then it is a homeomorphism._

(vii) _If_ $f : C \to H$ _as in (v) is a surjection then_ H _has the quotient topology._

Proof. (i) Let X be compact, and let $Y \subseteq X$ be closed; let $Y = \cup Y_i$, where each Y_i is open in Y; then $Y_i = Y \cap X_i$ for some open X_i in X. The X_i together with $X \backslash Y$ form an open cover of X; since X is compact this has a finite subcover, from which we may construct a finite cover of Y by deleting $X \backslash Y$ (if it appears) and replacing X_i by Y_i. This is the required finite subcover of $\{Y_i\}$.

(ii) Let H be a Hausdorff space, and let C be a compact subspace of H. Let $x \notin C$ be fixed. We have to find a neighbourhood X of x with $X \cap C = \emptyset$. For every $c \in C$ we can find an open neighbourhood N_c of c and an open neighbourhood X_c of x in H such that $N_c \cap X_c = \emptyset$. The sets N_c cover C, so by compactness there exist finitely many $c, d, \ldots, g \in C$ such that $C \subseteq N_c \cup N_d \cup \ldots \cup N_g$. Let $X = X_c \cap X_d \cap \ldots \cap X_g$; this is a neighbourhood of x, and $X \cap C = \emptyset$. Hence C is closed.

(iii) _Exercise_ (look at the pre-image of an open cover of the image space).

(iv) This is also a routine exercise.

(v) Let $f : C \to H$ be continuous with C compact and H Hausdorff. If $X \subseteq C$ is closed, it is compact by (i), so Xf is compact by (iii) and so closed by (ii).

(vi) From (v), f is closed, continuous and bijective; it is therefore a homeomorphism.

(vii) By Proposition 1 ($\mathcal{T}$), since f is continuous and closed, it induces a homeomorphism $C/(ff^{-1}) \cong H$.

Examples. (i) We know that any bounded closed set in $\mathbf{R}^n$ is compact. Any subset of $\mathbf{R}^n$ defined by a set of equations or inequalities of the form $f(x_1, \ldots, x_n) = \alpha$, or $\leq \alpha$, or $\geq \alpha$, where $f : \mathbf{R}^n \to \mathbf{R}$ is continuous and $\alpha \in \mathbf{R}$, is closed. In particular S^n is closed in $\mathbf{R}^{n+1}$ since it is defined by $x_0^2 + x_1^2 + \ldots + x_n^2 = 1$. It is also bounded, so it is compact.

(ii) Real projective n-space $\mathbf{P}^n$ is defined as $(\mathbf{R}^{n+1}\setminus\{0\})/\rho$ where $x\rho y$ means that $\exists\lambda\neq 0$ with $x = \lambda y$. The compact subspace S^n of $\mathbf{R}^{n+1}\setminus\{0\}$ maps onto $\mathbf{P}^n$ under the quotient map. Hence $\mathbf{P}^n$ is compact. ($\mathbf{P}^n$ is in fact a quotient space of S^n, obtained by identifying diametrically opposite points.)

(iii) $T^n \cong \mathbf{R}^n/\mathbf{Z}^n$ is compact because the closed cell given by $0 \leq x_i \leq 1$ in $\mathbf{R}^n$ is closed and bounded, so compact, and it maps onto T^n.

(iv) The orthogonal group O_n is the subspace of $\mathbf{R}^{n^2}$ defined by the equations

$$\sum_i x_{ij}x_{ik} = \delta_{jk} \quad \text{for all } j, k.$$

It is therefore closed in $\mathbf{R}^{n^2}$. It is also bounded since $\sum_i x_{ij}^2 = 1$ for all j. Hence O_n is a compact group. The subgroup O_n^+ is open, therefore closed; so O_n^+ is also a compact group for all $n \geq 1$.

We can now complete the proof that O_n^+ is connected (see Example (x) of §7 above). We there constructed a continuous map $\phi : O_{n+1}^+ \to S^n$, and we only need to show that ϕ is a closed map. But this now follows from Proposition 10 ($\mathcal{T}$), part (v), since O_{n+1}^+ is compact and S^n is a Hausdorff space. (A similar inductive argument can be applied to the connectedness of the unitary and symplectic groups; see Chevalley [2] for the details.)

Theorem 2 (Tychonoff's Theorem). Any product of compact spaces is compact.

We shall give a proof using terms familiar to algebraists. At the same time we shall describe the connection between ideals and filters so that the analytically minded reader can translate into more familiar terms; alternatively, he will find other proofs in [6] and [7].

Preliminaries. Let S be any set and let $\mathcal{L}$ be any lattice of subsets of S (i.e. $\mathcal{L}$ is closed with respect to $\cap$ and $\cup$, and $\emptyset \in \mathcal{L}$, $S \in \mathcal{L}$). An ideal of $\mathcal{L}$ is a subset $\mathcal{J} \subseteq \mathcal{L}$ such that

(i) if $A, B \in \mathcal{J}$ then $A \cup B \in \mathcal{J}$;

(ii) if $A \in \mathcal{J}$ and $B \subseteq A$, $B \in \mathcal{L}$ then $B \in \mathcal{J}$.

$\mathcal{J}$ is a _proper_ ideal if $\mathcal{J} \neq \mathcal{L}$, i. e., if $S \notin \mathcal{J}$.

The union of any chain of proper ideals is a proper ideal, so by Zorn's lemma every proper ideal is contained in a maximal ideal. ('Maximal ideal' will always mean 'maximal proper ideal'.)

Lemma 1. _If_ $\mathfrak{M}$ _is a maximal ideal of_ $\mathcal{L}$ _and if_ $A, B \in \mathcal{L}$ _and_ $A \cap B \in \mathfrak{M}$ _then either_ $A \in \mathfrak{M}$ _or_ $B \in \mathfrak{M}$.

Proof. We show that if $A \notin \mathfrak{M}$ then $B \in \mathfrak{M}$. If $A \notin \mathfrak{M}$, put

$$\mathcal{J} = \{X \in \mathcal{L}; X \subseteq A \cup M, \text{ for some } M \in \mathfrak{M}\}.$$

Then $\mathcal{J}$ is an ideal of $\mathcal{L}$. Since $A \in \mathcal{J}$ and $\mathfrak{M} \subseteq \mathcal{J}$, $\mathcal{J}$ properly contains $\mathfrak{M}$, whence $\mathcal{J} = \mathcal{L}$. In particular $S \in \mathcal{J}$, so $S = A \cup M$, for some $M \in \mathfrak{M}$. Hence $B \subseteq (A \cap B) \cup M$, and since $A \cap B \in \mathfrak{M}$ and $M \in \mathfrak{M}$, this implies that $B \in \mathfrak{M}$.

Dually, if $\mathcal{L}$ is any lattice of subsets, a _filter_ in $\mathcal{L}$ is a set $\mathfrak{F} \subseteq \mathcal{L}$ satisfying:

(i) $A, B \in \mathfrak{F} \Rightarrow A \cap B \in \mathfrak{F}$,

(ii) if $A \in \mathfrak{F}$, $B \in \mathcal{L}$, and $B \supseteq A$, then $B \in \mathfrak{F}$,

(iii) $\emptyset \notin \mathfrak{F}$.

By Zorn's Lemma, every filter is contained in a maximal one (maximal filters are called _ultrafilters_). By duality we clearly have:

Lemma 1'. _If_ $\mathfrak{F}$ _is an ultrafilter and_ $A \cup B \in \mathfrak{F}$, _then_ $A \in \mathfrak{F}$ _or_ $B \in \mathfrak{F}$.

Now let X be a topological space. We have two lattices of subsets of X, namely

$\mathcal{C}(X)$, the lattice of closed subsets, and

$\mathcal{O}(X)$, the lattice of open subsets.

These are dual-isomorphic (i. e. unions and intersections are interchanged) under the complement map: $A \mapsto X \backslash A$. In this correspondence, filters in $\mathcal{C}(X)$ correspond to proper ideals in $\mathcal{O}(X)$, and ultrafilters correspond to maximal ideals.

Lemma 2. <u>The following are equivalent conditions on a topological space</u> X:

(i) X <u>is compact.</u>

(ii) <u>If</u> $\mathcal{J}$ <u>is a proper ideal in</u> $\mathcal{O}(X)$, <u>then</u>

$\cup(\mathcal{J}) \neq X$.

(iii) <u>If</u> $\mathcal{F}$ <u>is a filter in</u> $\mathcal{C}(X)$, <u>then</u>

$\cap(\mathcal{F}) \neq \emptyset$.

Proof. (i) $\Rightarrow$ (ii): Let X be compact, and suppose that $\cup(\mathcal{J}) = X$. Then by compactness there are finitely many elements $A_1, \ldots, A_n \in \mathcal{J}$ whose union is X. But $\mathcal{J}$ is closed under finite unions, so $X \in \mathcal{J}$, and $\mathcal{J}$ is not a proper ideal.

(ii) $\Rightarrow$ (i): Assume (ii), and let $X = \cup X_i$ be any open cover of X. Let $\mathcal{J}$ be the collection of all open subsets A of X such that A is covered by finitely many of the X_i. Then $\mathcal{J}$ is an ideal in $\mathcal{O}(X)$, and each $X_i \in \mathcal{J}$, so $X \subseteq \cup(\mathcal{J})$. Hence, by assumption, $\mathcal{J}$ cannot be proper, so $X \in \mathcal{J}$. Thus any open cover of X can be reduced to a finite subcover, and X is compact.

(ii) $\Longleftrightarrow$ (iii) by taking complements.

Proof of Tychonoff's Theorem. We phrase the proof in terms of ideals, but it can equally be written in terms of filters.

Let I be a set, and for each $i \in I$ let X_i be a compact topological space. Let $X = \Pi X_i$, and suppose that $\mathcal{J}$ is a proper ideal in $\mathcal{O}(X)$. We show that $\cup(\mathcal{J}) \neq X$, and then by Lemma 2 we see that X is compact.

By Zorn's Lemma, $\mathcal{J}$ is contained in some maximal ideal $\mathcal{M}$. It is therefore enough to show that $\cup(\mathcal{M}) \neq X$ for an arbitrary maximal ideal $\mathcal{M}$.

For each $i \in I$, define

$$\mathcal{M}_i = \{A \in \mathcal{O}(X_i);\ A\pi_i^{-1} \in \mathcal{M}\},$$

where π_i is the projection $X \to X_i$. This is clearly a proper ideal of

$\mathcal{O}(X_i)$, hence as X_i is compact, $\cup(\mathfrak{M}_i) \neq X_i$. Choose $x_i \in X_i \backslash \cup(\mathfrak{M}_i)$. We claim that $x = (x_i) \notin \cup(\mathfrak{M})$. For if $x \in A \in \mathfrak{M}$, A is open, so there is a basic open set Y such that $x \in Y \subseteq A$. (Hence $Y \in \mathfrak{M}$.) Now we may choose finitely many indices $i(1), \ldots i(m) \in I$, and open sets

$$Z_{i(r)} \subseteq X_{i(r)} \qquad (r = 1, \ldots m)$$

such that, if $Y_{(r)} = Z_{i(r)}\pi_{i(r)}^{-1}$, we have

$$Y = Y_{(1)} \cap \ldots \cap Y_{(m)}.$$

Then for each r, $x \in Y \subseteq Y_{(r)}$, and therefore $x_{i(r)} \in Z_{i(r)}$. But by Lemma 1, since $Y \in \mathfrak{M}$, some $Y_{(r)} \in \mathfrak{M}$, and then $x_{i(r)} \in Z_{i(r)} \in \mathfrak{M}_{i(r)}$, which contradicts our choice of the x_i. Hence $\cup(\mathfrak{M}) \neq X$, and X is compact.

The converse of Tychonoff's Theorem is trivial. If each X_i is non-empty, and ΠX_i is compact, then each projection is continuous, hence its image is also compact; but this is the whole of X_i.

Proposition 10 ($\mathcal{T}\mathcal{G}$). _Let_ G, $G_i \in \mathcal{T}\mathcal{G}$, _and let_ H _be a subgroup of_ G. _Then_

(i) G _compact and_ H _closed_ $\Rightarrow$ H _compact._

(ii) G _compact_ $\Rightarrow$ G/H _compact._

(iii) H _compact and_ G/H _compact_ $\Rightarrow$ G _compact._

(iv) ΠG_i _compact_ $\Leftrightarrow$ _each_ G_i _compact._

Proof. (i), (ii), (iv) follow from Proposition 10 ($\mathcal{T}$) and Theorem 2. For the proof of (iii) we need:

Proposition 11. _Let_ $G \in \mathcal{T}\mathcal{G}$, _let_ C _be a compact subset of_ G _and let_ A _be an open set containing_ C. _(We call such a set an open neighbourhood of_ C.) _Then there is an open neighbourhood_ V _of_ e _in_ G _such that_ $VC \subseteq A$.

Proof. Every point $x \in C$ is an interior point of A, so there is an open neighbourhood W_x of e such that $W_x x \subseteq A$. Choose an open neighbourhood V_x of e such that $V_x V_x \subseteq W_x$. The open sets $V_x x$

cover C, hence so do a finite number; thus we may select $V_1, \ldots, V_n$, corresponding to $x_1, \ldots, x_n$, with $C \subseteq \bigcup_{i=1}^{n} V_i x_i$.

Now take $V = \cap V_i$, an open neighbourhood of e. Then $VC \subseteq VV_1x_1 \cup \ldots \cup VV_nx_n$, and each $VV_i \subseteq V_iV_i \subseteq W_i$, so $VV_ix_i \subseteq A$. Thus $VC \subseteq A$ as required.

Proof of (iii) in Proposition 10 ($\mathcal{TG}$) above. Suppose that H and G/H are compact. Write q for the projection $G \rightarrow G/H$. Let $\mathcal{J}$ be an ideal of $\mathcal{O}(G)$ such that $\cup(\mathcal{J}) = G$. It is enough to show that $G \in \mathcal{J}$.

For any $x \in G$, xH is compact, and is covered by $\mathcal{J}$ (as $\cup(\mathcal{J})=G$); hence $xH \subseteq J$, where J is a finite union of sets in $\mathcal{J}$. But $\mathcal{J}$ is an ideal, so $J \in \mathcal{J}$, and J is open.

By Proposition 11 we may choose an open neighbourhood V of e such that $VxH \subseteq J$. Then VxH is open, hence is a member of $\mathcal{J}$, and it is a union of cosets of H.

Let $\mathcal{J}' = \{A \in \mathcal{O}(G/H) : Aq^{-1} \in \mathcal{J}\}$. Then $\mathcal{J}'$ is an ideal in $\mathcal{O}(G/H)$, non-empty since

$$(VxH)qq^{-1} = VxH \in \mathcal{J} \Rightarrow (VxH)q \in \mathcal{J}'.$$

But $(VxH)q$ contains $(exe)q = xq$, so for any x, $xq \in \cup(\mathcal{J}')$ and $\cup(\mathcal{J}') = G/H$. Since G/H is compact, $G/H \in \mathcal{J}'$, i.e., $G \in \mathcal{J}$, and G is therefore compact.

Proposition 12. <u>In a compact group</u> G, <u>every open subgroup</u> H <u>has finite index.</u>

Proof. G is a union of disjoint cosets of H, which are open; hence it is a union of finitely many of them. (Alternatively, G/H is compact (Proposition 10) and discrete (Proposition 6), hence finite.)

Proposition 13. <u>In a Hausdorff space</u> H, <u>any two disjoint compact sets</u> A <u>and</u> B <u>have disjoint open neighbourhoods</u> A', B' <u>in</u> H <u>(i.e., there exist disjoint open sets</u> A', B' <u>with</u> $A' \supseteq A$, $B' \supseteq B$).

Proof. Fix $a \in A$. For each $b \in B$ there are disjoint open neighbourhoods U_b of a, V_b of b. The V_b cover B, hence so do a finite number of them, say $V_1, \ldots, V_n$, corresponding to $b_1, \ldots, b_n$ and $U_1, \ldots, U_n$. Let $V = \bigcup_{i=1}^{n} V_i$, $U = \bigcap_{i=1}^{n} U_i$. Then U, V are disjoint open sets with $a \in U$, $B \subseteq V$, by construction.

Now, for each $a \in A$, let S_a, T_a be disjoint open sets such that $a \in S_a$, $B \subseteq T_a$. (We have just proved that such sets exist.) The S_a cover A, hence so do a finite number, say $S_1, \ldots, S_m$, corresponding to $a_1, \ldots, a_m$ and $T_1, \ldots, T_m$. Then $A' = \bigcup_{j=1}^{m} S_j$, and $B' = \bigcap_{j=1}^{m} T_j$ are disjoint open sets with $A \subseteq A'$, $B \subseteq B'$, as required.

Proposition 14. In a compact Hausdorff space H, the component at any point x is the intersection of all open-closed sets containing x.

Proof. Let C_i $(i \in I)$ be all the open-closed sets containing x, and let C be their intersection. Let C_i^*, C^* be the complements in H of C_i, C, respectively. Each C_i contains comp(x), as the latter is connected; hence $C \supseteq \text{comp}(x)$. It is therefore sufficient to show that C is connected (because this implies $C \subseteq \text{comp}(x)$).

Suppose that $C = X \cup Y$, where X, Y are disjoint sets, open-closed relative to C, with $x \in X$. It is enough to show that $X = C$. Now X, Y are closed in C, which is closed in H, so X, Y are closed in the compact space H, and are therefore compact. Hence by Proposition 13 there are disjoint open subsets X', Y' of H, with $X' \supseteq X$, $Y' \supseteq Y$.

Now $X' \cup Y' \supseteq C$, and $C^* = \cup C_i^*$, so H is covered by $\{X', Y', C_i^*\}$. Hence by compactness H can be written as a finite union, say

$$\begin{aligned} H &= C_1^* \cup C_2^* \cup \ldots \cup C_n^* \cup X' \cup Y' \\ &= D^* \cup X' \cup Y', \end{aligned} \tag{1}$$

where $D = C_1 \cap C_2 \cap \ldots \cap C_n$ is an open-closed neighbourhood of x and therefore contains C. It follows (using (1)) that $C \subseteq D \subseteq X' \cup Y'$. If we write $A = D \cap X'$, $B = D \cap Y'$, we deduce that $D = A \cup B$, with A, B disjoint and open-closed in D, which is open-closed in H, whence A, B are open-closed in H. But $x \in A$, so $C \subseteq A$, by definition of C, and we now have $C = C \cap A = C \cap D \cap X' = C \cap X' = X$, as required.

9. Profinite groups

Inverse limits. Let $\mathcal{C}$ be any category, and $(I, \leq)$ a partially ordered set. Suppose that for each $i \in I$ we have an object A_i in $\mathcal{C}$, and that for each pair $i \leq j$ we have a morphism $f_{ji} : A_j \to A_i$ in $\mathcal{C}$ satisfying the coherence conditions:

(i) each $f_{ii} : A_i \to A_i$ is the identity,

(ii) if $i \leq j \leq k$, then $f_{kj} f_{ji} = f_{ki} : A_k \to A_i$.

We then call $\{A_i, f_{ji}\}$ an inverse system in $\mathcal{C}$.

Let $A \in \mathcal{C}$. If $S = \{A_i, f_{ji}\}$ is an inverse system in $\mathcal{C}$, and if for each i, $\sigma_i : A \to A_i$ is a morphism in $\mathcal{C}$ such that whenever $i \leq j$ we have $\sigma_j f_{ji} = \sigma_i$, we shall call $\{A, \sigma_i\}$ a cone over S. If also $\{B, \tau_i\}$ is a cone over S, a morphism of cones

$$\theta : \{B, \tau_i\} \to \{A, \sigma_i\}$$

is a morphism $\theta : B \to A$ in $\mathcal{C}$ such that for every i, $\theta\sigma_i = \tau_i$.

If $\{A, \sigma_i\}$ is universal amongst cones over S, i.e. if for every cone $\{B, \tau_i\}$ over S there is a unique morphism of cones $\theta : \{B, \tau_i\} \to \{A, \sigma_i\}$, we shall call $\{A, \sigma_i\}$ the inverse (or projective) limit of S. If such an A exists, by the standard argument it is unique up to isomorphism (in fact $\{A, \sigma_i\}$ is unique up to isomorphism of cones), and we shall write

$$A = \varprojlim A_i \text{ in this case.}$$

Example. If I is trivially ordered ($i \leq j$ if and only if $i = j$), then A is the product in $\mathcal{C}$ of the A_i, and the morphisms $\sigma_i : A \to A_i$ are the projections.

In the category $\mathcal{TG}$, inverse limits always exist.

Proof. (This also works in $\mathcal{S}$, $\mathcal{T}$, $\mathcal{G}$.) Let $\{A_i, f_{ji}\}$ be an inverse system in $\mathcal{TG}$. Let $P = \Pi A_i$ in $\mathcal{TG}$, with projections $\pi_i : P \to A_i$.

For each $j \leq k$, the difference kernel of $\pi_k f_{kj}$ and $\pi_j : P \to A_j$ is a subgroup of P. Let A denote the intersection of all these subgroups, and let $\sigma_i : A \to A_i$ be the maps induced by the π_i. Then clearly A is a topological group, the σ_i are continuous group-homomorphisms, and by construction, if $j \leq k$, then $\sigma_k f_{kj} = \sigma_j$. In fact

$$A = \{a = \{a_i\} \in P;\ \text{if}\ j \le k,\ a_k f_{kj} = a_j\}.$$

Now, given morphisms $\tau_i : B \to A_i$ in $\mathcal{TG}$ such that, whenever $j \le k$, $\tau_k f_{kj} = \tau_j$, let $\tau : B \to P$ be the induced map, $b \mapsto b\tau = \{b\tau_i\}$. Then clearly $B\tau \subseteq A$, and the universal property holds. (The uniqueness of $\tau' : B \to A$ subject to $\tau'\sigma_i = \tau_i$ follows from the uniqueness of $\tau : B \to P$ subject to $\tau\pi_i = \tau_i$.) Hence $A = \varprojlim A_i$ in $\mathcal{TG}$.

Proposition 15. Let $G = \varprojlim G_i$ in $\mathcal{TG}$. Then

(i) if all the G_i are Hausdorff, so is G;

(ii) if all the G_i are compact Hausdorff, so is G;

(iii) if all the G_i are totally disconnected, so is G.

Proof. (i), (iii): If the G_i are Hausdorff (resp. totally disconnected), then so is ΠG_i by Proposition 5 (resp. Proposition 9), whence so is the subgroup $G \subseteq \Pi G_i$.

(ii) If all the G_i are compact Hausdorff, then so is ΠG_i, by Proposition 5 and Theorem 2. We claim that G is a closed subgroup of $P = \Pi G_i$, hence compact. For G_i is Hausdorff, so the difference kernel of two continuous maps from any space into G_i is closed by Proposition 3. Hence by the construction of inverse limits, G is the intersection of a family of closed subgroups of P, and so is closed.

Definition. $G \in \mathcal{TG}$ is a profinite group, if it is of the form $\varprojlim G_i$ in $\mathcal{TG}$ for some inverse system of finite discrete groups G_i.

Theorem 3. Let $G \in \mathcal{TG}$. Then G is profinite if and only if it is compact and totally disconnected.

Proof. One way round is immediate, using Proposition 15, since finite discrete groups are compact Hausdorff and totally disconnected.

Conversely, let G be compact and totally disconnected. Since all components, i.e., all points of G, are closed, G is T_1, and so in fact Hausdorff (Proposition 3 ($\mathcal{TG}$)). By Proposition 14, $\{e\} = \text{comp}(e)$ is the intersection of all open-closed neighbourhoods of e. We claim that every open-closed neighbourhood of e contains an open (and so closed) normal subgroup. To prove this, let A be any open-closed set with

$e \in A$. Then (as G is compact Hausdorff), A is compact, and open. Hence by Proposition 11 there is an open set V containing e, such that $VA \subseteq A$.

Let $W = V \cap V^{-1}$, so that W is open, $e \in W$, and $W = W^{-1}$. Then $WA = W^{-1}A \subseteq A$, and so by induction, $W^n A \subseteq A$ for all integers n. Since $e \in A$, it follows that $W^n \subseteq A$, for all integers n. If H is the subgroup generated by W, then $H = \bigcup_{n \in \mathbf{Z}} W^n \subseteq A$, and H is open by Proposition 6(ii). Since H is open in the compact group G, it has finite index (Proposition 12) and so has only finitely many distinct conjugates $x^{-1}Hx$ in G. The intersection K of these is thus an open normal subgroup of G, and $K \subseteq H \subseteq A$, as claimed.

Let $N_i (i \in I)$ be all the distinct open, normal subgroups of G. The argument above shows that $\bigcap_{i \in I} N_i = \{e\}$. Since G is compact, and N_i is open, the index of N_i in G is finite for all i (Proposition 12). Now order I by letting $i \leq j$ if and only if $N_i \supseteq N_j$, and put $A_i = G/N_i$, a finite discrete group. If $i \leq j$ then there exists a canonical morphism $f_{ji} : A_j \to A_i$ given by $xN_j \mapsto xN_i$. The $\{A_i\}$ and $\{f_{ji}\}$ form an inverse system of topological groups. Let $A = \varprojlim A_i$ in $\mathcal{TG}$ with canonical maps $\sigma_i : A \to A_i$. Each A_i is finite, so A is a profinite group; it is constructed as the subgroup of ΠA_i consisting of those elements $\underline{a} = \{a_i\} = \{x_i N_i\}_{i \in I}$ such that $x_j N_j \subseteq x_i N_i$, whenever $i \leq j$. Now G has quotient maps $\tau_i : G \to A_i$ for all i, and $\tau_j f_{ji} = \tau_i$ for $i \leq j$. Therefore there exists a unique $\tau : G \to A$ in $\mathcal{TG}$ such that $\tau\sigma_i = \tau_i$ for all $i \in I$, namely $x \to \{xN_i\}_{i \in I}$. We show:

(i) τ <u>is an injection.</u> For if $x \in G$ and $x\tau = e$, then $x\tau_i = e$ for all i, whence $x \in \bigcap N_i = \{e\}$.

(ii) τ <u>is a surjection.</u> Let $\underline{a} = \{x_i N_i\}_{i \in I}$ be any element of A. Each N_i is open and hence closed, so each $x_i N_i$ is closed. The intersection of any finite number of the N_i's, say $N_{i_1} \cap N_{i_2} \cap \ldots \cap N_{i_r}$, is again an open normal subgroup N_k for some k. Hence $x_k N_k \subseteq x_{i_\alpha} N_{i_\alpha}$ for $\alpha = 1, \ldots, r$, and so the intersection of these cosets is non-empty. By the compactness of G there exists $x \in \bigcap_{i \in I} x_i N_i$. But then $xN_i = x_i N_i$

for all $i \in I$ and so $x\tau = \underline{a}$.

We have now proved that τ is a continuous bijection $G \to A$. However G is compact and A is Hausdorff. By Proposition 10, τ is an isomorphism in $\mathcal{TG}$ and so G is a profinite group.

Theorem 4. *Let G be any compact group and let G_0 be the connected component of G at e. Then*

(i) *G_0 is a connected compact group;*

(ii) *G/G_0 is a profinite group;*

(iii) *G_0 is the intersection of all open, normal subgroups of G.*

Proof. Put together Theorems 1 and 3.

Example. (i) Consider $\mathbf{Z}$ with the p-adic topology. The subgroups $p^n\mathbf{Z}$ form a fundamental system of open neighbourhoods of 0. Now $\cap\, p^n\mathbf{Z} = \{0\}$ and so $\mathbf{Z}$ is totally disconnected. However $\mathbf{Z}$ is not compact because, e.g., the cosets $p\mathbf{Z}+1 \supseteq p^2\mathbf{Z}+p+1 \supseteq p^3\mathbf{Z}+p^2+p+1 \supseteq \ldots$ have empty intersection (when $p \neq 2$). In other words the congruences

$$(1 - p)x \equiv 1 \pmod{p^n}$$

have no common solution x for all n, even though every finite number of them have a common solution. (*Exercise*: show that $\mathbf{Z}$ is not compact with respect to the 2-adic topology.)

However, if we form the p-*adic integers* $\mathbf{Z}_p = \varprojlim (\mathbf{Z}/p^n\mathbf{Z})$ we obtain a compact group containing $\mathbf{Z}$ as a subgroup (exercise). In $\mathbf{Z}_p$ the above congruences do have a common solution, and this was the original purpose of Hensel's construction of the p-adic integers.

Exercise. Every element of $\mathbf{Z}_p$ is uniquely expressible in the form

$$b_0 + b_1 p + b_2 p^2 + \ldots \qquad \text{(infinite sum)}$$

where $b_i \in \mathbf{Z}$, and $0 \le b_i < p$. (Infinite sums are defined as limits of finite sums in the usual way, relative to the p-adic topology.)

Example. (ii) Infinite Galois groups. Let K be any field, L any Galois extension of K (i.e., algebraic, normal and separable), not necessarily of finite degree.

Let $G = \text{Gal}(L : K)$ be the group of automorphisms of L fixing all elements of K. Clearly L is generated as a field by all finite Galois extensions L_i of K with $L_i \subseteq L$. For each L_i the subgroup $N_i = \text{Gal}(L : L_i)$ of G is normal and of finite index in G. Furthermore $G/N_i \cong \text{Gal}(L_i : K)$ in $\mathcal{G}$.

Define a topology on G by taking $\{N_i\}$ as a fundamental system of neighbourhoods of e. (This may be done since $N_i \cap N_j = N_k$, where L_k is the subfield of L generated by L_i and L_j. See §4 above.)

It is not difficult to show that there exists a one-to-one correspondence between the closed subgroups of G, with respect to this topology, and the fields between K and L; this correspondence has the properties one expects from the Galois theory of finite extensions.

Exercises. (i) $G = \text{Gal}(L : K)$ is profinite; $G = \varprojlim G/N_i$. Crucial points in the proof are:

(a) If $\sigma_i \in \text{Gal}(L_i : K)$ for each i, and σ_i, σ_j agree on $L_i \cap L_j$ for all i and j, then $\exists \sigma \in G$ inducing all σ_i.

(b) The topology on G is the same as the inverse limit topology.

(ii) If G is in $\mathcal{TG}$, and if A and B are compact subsets of G, then AB is compact.

(iii) If A is a closed subset of the topological group G, and B is a compact subset of G, then AB is closed.

(iv) If H is a compact subgroup of the topological group G, then the quotient map $G \to G/H$ is a closed map.

10. Locally compact groups

Definition. A topological space X is locally compact if every point of X has a compact neighbourhood.

(Note: Some authors require that every point has a closed, compact neighbourhood. The two definitions coincide if X is Hausdorff, or if X is regular. In particular, they coincide for topological groups, by Proposition 4.)

Examples. (i) $\mathbf{R}^n$ is locally compact.

(ii) Any compact space is locally compact.

(iii) Any discrete space is locally compact.

(iv) $\mathbf{Q} \subseteq \mathbf{R}$ is not locally compact.

(v) Infinite-dimensional Hilbert space is not locally compact.

(vi) An infinite product of copies of $\mathbf{R}$ is not locally compact (as is implied by the next proposition).

Proposition 16 ($\mathcal{T}$). (i) Any closed subspace of a locally compact space is locally compact.

(ii) If X is locally compact, and $f : X \to Y$ is a continuous, open surjection, then Y is locally compact.

(iii) If X_i is a non-empty topological space for every $i \in I$, then $\prod_{i \in I} X_i$ is locally compact if and only if each X_i is locally compact and all but a finite number are compact.

Proof. (i) Let X be locally compact, let Y be a closed subspace of X, and let $y \in Y$. There exists a compact neighbourhood N of y in X. Now $N \cap Y$ is a neighbourhood of y in Y, and it is compact since it is closed in N.

(ii) Let $y \in Y$; then since f is surjective, $y = xf$ for some $x \in X$. There exists a compact neighbourhood N of x in X. Since f is continuous and open, Nf is a compact neighbourhood of y in Y.

(iii) Suppose the X_i are locally compact and all but a finite number compact. Let $\underline{x} \in \Pi X_i = X$, where $\underline{x} = \{x_i\}$, $x_i \in X_i$. There exists for each i a compact neighbourhood N_i of x_i in X_i. For all but a finite number of i we may take $N_i = X_i$. Hence ΠN_i is a compact neighbourhood of $\underline{x}$ in X. Conversely, suppose that $X_i \neq \emptyset$ for all i, and that $X = \Pi X_i$ is locally compact. The projections $\pi_i : X \to X_i$ are continuous, open surjections and so each X_i is locally compact by (ii). Now there exists an $\underline{x} \in X$ and a compact neighbourhood N of $\underline{x}$ in X. For all but a finite number of i we have $N\pi_i = X_i$, so almost all the X_i are compact.

Note: The image of a locally compact space under a continuous map is not, in general, locally compact. (For any X, let X_0 be X

with the discrete topology. Then X_0 is locally compact and the identity map $X_0 \to X$ is continuous.)

Proposition 16 ($\mathcal{T}\mathcal{G}$). Let G, G_i be topological groups and let H be a subgroup of G. Then:

(i) G is locally compact if and only if there exists a compact neighbourhood of e;

(ii) G locally compact and H closed in $G \Rightarrow H$ locally compact;

(iii) G locally compact $\Rightarrow G/H$ locally compact;

(iv) H locally compact and G/H locally compact $\Rightarrow G$ locally compact;

(v) ΠG_i is locally compact if and only if all the G_i are locally compact and all but a finite number are compact.

Proof. (i): G is homogeneous.

(ii) and (v) follow directly from Proposition 16 ($\mathcal{T}$), whilst (iii) needs only the additional information that the quotient map $G \to G/H$ is a continuous, open, surjection.

(iv): It is enough to find closed neighbourhoods U and V of e in G with the following properties:

(a) $V^{-1}V \subseteq U$ (therefore $V \subseteq U$);

(b) $xH \cap U$ is compact for all $x \in U$ (therefore $xH \cap V$ is compact);

(c) if $q : G \to G/H$ is the quotient map, then $C = Vq$ is a compact subspace of G/H.

Given such a U and V we argue as in Proposition 10 ($\mathcal{T}\mathcal{G}$) (iii) to show that V is compact. The argument is, however, a little more difficult. Let $\mathcal{J}$ be an ideal of $\mathcal{O}(V)$ and suppose that $\cup(\mathcal{J}) = V$. Let $\mathcal{J}' = \{A \in \mathcal{O}(C) : Aq^{-1} \cap V \in \mathcal{J}\}$. We shall deduce that $\cup(\mathcal{J}') = C$.

Let $x \in V$; then $xH \cap V$ is compact and is covered by elements of $\mathcal{J}$. Hence there exists $J \in \mathcal{J}$ such that $J \supseteq xH \cap V$. Now $xH \cap U \subseteq J \cup (G\backslash V) = X$, which is an open set of G. By Proposition 11, there exists an open neighbourhood W of e in G such that $W(xH \cap U) \subseteq X$. We may in fact choose W to be a subset of V. Now $WxH \cap V \subseteq J$; for suppose $wxh = v$ where $w \in W$, $h \in H$ and $v \in V$. Then $xh = w^{-1}v \in V^{-1}V \subseteq U$. Hence $v = wxh \in W(xH \cap U) \subseteq X$. Therefore $v \in X \cap V = J$. It follows that $WxH \cap V \in \mathcal{J}$.

Now Wx is open in G (because W is open), so $(Wx)q$ is open in G/H. Therefore $(Wx)q \cap C \in \mathcal{J}'$ since $Wxqq^{-1} \cap V = WxH \cap V \in \mathcal{J}$. In particular $xq \in \cup(\mathcal{J}')$ for any $x \in V$, so $\cup(\mathcal{J}') = C$. But C is compact so $C \in \mathcal{J}'$. Therefore $V \in \mathcal{J}$ and by Lemma 2 (p. 47) V is compact.

It remains to find closed neighbourhoods U and V of e satisfying conditions (a), (b), (c). The subgroup H of G is locally compact; therefore there exists a neighbourhood U_0 of e in G such that $H \cap U_0$ is compact. Since G is regular (by Proposition 4) there exists a closed neighbourhood U_1 of e in G with $U_1 \subseteq U_0$. Since $U_1 \cap H$ is closed in $U_0 \cap H$ it follows that $U_1 \cap H$ is compact. Again by regularity there exists a closed neighbourhood U of e with $U^{-1}U \subseteq U_1$. Now if $x \in U$ then $xH \cap U = x(H \cap x^{-1}U)$. Since $x^{-1}U$ is closed and contained in U_1, it follows that $H \cap x^{-1}U$ is closed in $H \cap U_1$ and is therefore compact. Hence $xH \cap U$ is compact.

Now choose a neighbourhood V_0 of e in G such that $V_0^{-1}V_0 \subseteq U$. Since q is an open map, V_0q is a neighbourhood of H in G/H. Using the regularity of G/H (Proposition 4(ii)), there exists a closed, compact, neighbourhood C of H in G/H with $C \subseteq V_0q$.

Now put $V = V_0 \cap Cq^{-1}$. It is easy to see that U and V satisfy conditions (a), (b), (c).

Proposition 17. *Let* X *be a locally compact Hausdorff space. Let* C *be a compact subspace of* X. *Then every neighbourhood of* C *in* X *contains a compact neighbourhood of* C; *i. e., given* $N \supseteq U \supseteq C$ *with* U *open, we can find compact* C' *and open* U' *such that* $N \supseteq C' \supseteq U' \supseteq C$.

Proof. We first suppose that X is compact. Let N be a neighbourhood of C; we may assume that N is open. Then $X \backslash N$ is compact. By Proposition 13 there exist open sets $V \supseteq C$, $W \supseteq X \backslash N$ with $V \cap U = \emptyset$. Put $C' = X \backslash W$; this is the required compact neighbourhood of C inside N.

Now consider the general case. Each point $c \in C$ has a compact neighbourhood N_c, whose interior we denote by M_c. Since C is compact, it is covered by a finite number of the M_c, say $C \subseteq M_{c_1} \cup M_{c_2} \cup \ldots \cup M_{c_n}$. If we now put $X_0 = N_{c_1} \cup N_{c_2} \cup \ldots \cup N_{c_n}$, then by Proposition 10, X_0 is compact and it is a neighbourhood of C. Given a neighbourhood N of C, let $N_0 = N \cap X_0$ and apply the special case above: there exists C', a compact neighbourhood of C in X_0, with $C' \subseteq N_0$. Since X_0 is a neighbourhood of C in X and C' is a neighbourhood of C in X_0 it follows that C' is a neighbourhood of C in X.

Corollary. Any open subspace of a locally compact Hausdorff space is locally compact.

Exercise. A subspace Y of X is said to be locally closed if every point $y \in Y$ has a neighbourhood N in X such that $N \cap Y$ is closed in N. Prove the following:

(i) Closed sets and open sets are locally closed.

(ii) In any space X, a locally closed subspace is a closed subspace of an open subspace, and conversely.

(iii) In a locally compact Hausdorff space X, the subspace Y is locally compact if and only if it is locally closed. (See Nachbin [7], p. 10, in case of difficulty.)

Examples. The following are locally compact groups.

1. (**R**, +).
2. Any discrete group.
3. Any compact group.
4. $GL_n(\mathbf{R})$ (since it is an open subspace of $\mathbf{R}^{n^2}$).
5. $SL_n(\mathbf{R})$ (since it is a closed subspace of $GL_n(\mathbf{R})$).
6. $PGL_n(\mathbf{R})$ (by Proposition 16 since it is a quotient group of $GL_n(\mathbf{R})$).

The corresponding complex groups are also locally compact.

(**Exercise.** $SL_n(\mathbf{R})$ is not compact if $n \geq 2$.)

7. $(\mathbf{R}^{pos}, \cdot)$, $(\mathbf{C}^*, \cdot)$.
8. The 'affine group': this is the group of transformations $\mathbf{R}^n \to \mathbf{R}^n$ generated by all linear isomorphisms and all translations. The

translations form a normal subgroup isomorphic with $\mathbf{R}^n$, whose quotient group is isomorphic with $GL_n(\mathbf{R})$; both these groups are locally compact, so the whole affine group is locally compact (by Proposition 16).

9. The additive group $\mathbf{Q}_p$ of p-adic numbers is locally compact. This group can be defined as $\varprojlim (\mathbf{Q}/p^n U)$, where U is the group of rational numbers with denominator prime to p. It contains $\mathbf{Z}_p = \varprojlim (\mathbf{Z}/p^n \mathbf{Z})$ as a subgroup, and in fact $\mathbf{Z}_p$ is an open, compact neighbourhood of 0 in $\mathbf{Q}_p$. (Exercise.)

III · Integration on locally compact groups

1. Abstract integrals

We seek to construct an 'integral' on a locally compact group. This may be done by a variety of methods; we wish to avoid measure-theoretic methods and so we take as our model the construction of the Riemann integral on $(\mathbf{R}, +)$. There we have the set $\mathcal{R}$ of all Riemann-integrable functions (vanishing outside some interval), and a map $\int : \mathcal{R} \to \mathbf{R}$ defined by

$$\int f = \int_{-\infty}^{\infty} f(x)dx$$

which satisfies:

(i) Linearity: $\int(\lambda f + \mu g) = \lambda \int f + \mu \int g$.

(ii) Positivity: $f \geq 0 \Rightarrow \int f \geq 0$.

(iii) Translation-invariance: if for some fixed $a \in \mathbf{R}$ and all $x \in \mathbf{R}$, $g(x) = f(x + a)$ then

$$\int g = \int f.$$

These three properties actually characterize the Riemann integral up to a scalar multiple. (Exercise.) The same is true if we consider only continuous functions on $\mathbf{R}$, and this suggests that we should try to construct, for more general groups, an integral with similar properties for real-valued continuous functions.

From now on we shall only consider locally compact spaces and groups, and we shall always assume that they are Hausdorff.

For a space X, $\mathcal{K}(X)$ denotes the set of all continuous functions $X \to \mathbf{R}$ with compact support, where

$$\text{support}(f) = \text{closure of } \{x \in X ; f(x) \neq 0\}.$$

(The reader will notice that we have reverted to a left-handed notation for functions $X \to \mathbf{R}$. This notation is more familiar in such a context and will cause no conflict since we shall not need to compose such functions.) Equivalently, $\mathcal{K}(X)$ is the set of continuous functions $X \to \mathbf{R}$ which vanish outside some compact set.

$\mathcal{K}(X)$ is a real vector space. It is also partially ordered by $\leq$, where $f \leq g$ means that $f(x) \leq g(x)$ for all x. In fact it is a lattice:

$$f, g \in \mathcal{K}(X) \Rightarrow \sup(f, g), \quad \inf(f, g) \in \mathcal{K}(X) \quad \text{(exercise)},$$

(but it has no largest or smallest member). In particular, since $|f| = \sup(f, -f)$,

$$f \in \mathcal{K}(X) \Rightarrow |f| \in \mathcal{K}(X).$$

Denote by $\mathcal{K}_+(X)$ the set of all $f \in \mathcal{K}(X)$ satisfying $f > 0$, i.e., $f \geq 0$ and $f \neq 0$ (in other words, $f(x) \geq 0$ everywhere, $f(x) > 0$ somewhere!). Then $\mathcal{K}_+(X)$ is a convex cone in $\mathcal{K}(X)$ (see p. 70 for the definition of convex cone).

It is by no means clear that $\mathcal{K}(X)$ is not the trivial vector space, so we hasten to establish

Proposition 18 (Urysohn's lemma). _Let_ X _be a non-empty locally compact Hausdorff space. Let_ $C \subseteq U \subseteq X$, _with_ C _compact and_ U _open. Then there exists a continuous function_ $f : X \to \mathbf{R}$ _such that_

(i) $f(x) = 1$ _for all_ $x \in C$,

(ii) $f(x) = 0$ _for all_ $x \notin U$,

(iii) $0 \leq f(x) \leq 1$ _for all_ $x \in X$,

(iv) $\text{support}(f)$ _is compact and contained in_ U.

Proof. By Proposition 17 there exists a compact neighbourhood $C(0)$ of C inside U. Write $C(1) = C$. We observe that, by Proposition 17, if C'' is a compact neighbourhood of the compact set C', then there exists a compact neighbourhood C''' of C' lying inside C'', such that C'' is a compact neighbourhood of C'''. We shall then say that C''' _separates_ C' and C''.

Choose $C(1/2)$ separating $C(0)$ and $C(1)$; then choose $C(1/4)$ separating $C(0)$ and $C(1/2)$, and $C(3/4)$ separating $C(1/2)$ and $C(1)$.

Continuing in this fashion we get $C(\theta)$ for every θ of the form $\theta = i/2^n$ with $0 \le i \le 2^n$. Then for any real α, with $0 \le \alpha \le 1$, define

$$C(\alpha) = \bigcap_{\theta \le \alpha} C(\theta) \quad (\text{where } \theta \text{ is of the form } i/2^n).$$

Since each $C(\theta)$ is closed, $C(\alpha)$ itself is closed, and so compact. Define $C(\alpha) = \mathbf{X}$ if $\alpha < 0$, $C(\alpha) = \emptyset$ if $\alpha > 1$. Then for any $\alpha < \beta$, $C(\alpha)$ is a closed neighbourhood of $C(\beta)$, compact if $\alpha \ge 0$. Now define $f(x) = \sup\{\alpha ; x \in C(\alpha)\}$. Immediately we see that $f(x) = 0$ for all x outside $C(0)$, so that support(f) is compact and contained in U. Furthermore, $f(x) = 1$ for all $x \in C(1)$, and $0 \le f(x) \le 1$ everywhere.

To complete the proof, we need to show that f is continuous. For real β and γ we observe that on the one hand

$$f(x) \ge \beta \Longleftrightarrow x \in \bigcap_{\alpha < \beta} C(\alpha) \quad = D(\beta),$$

and $D(\beta)$ is a closed set; while on the other hand

$$\begin{aligned} f(x) > \gamma &\Longleftrightarrow x \in C(\alpha) \text{ for some } \alpha > \gamma \\ &\Longleftrightarrow x \in \text{interior of } C(\alpha') \text{ for some } \alpha' > \gamma \\ &\Longleftrightarrow x \in \bigcup_{\alpha' > \gamma} \operatorname{Int} C(\alpha') \quad = E(\gamma), \end{aligned}$$

and $E(\gamma)$ is an open set. Thus

$$\gamma < f(x) < \beta \Longleftrightarrow x \in E(\gamma) \cap (\mathbf{X} \backslash D(\beta)),$$

which is open; so the pre-image of a basic open set in $\mathbf{R}$ is open in X, and f is continuous.

Corollary. If X is a non-empty locally compact Hausdorff space, then

$$\mathcal{K}(\mathbf{X}) \ne \{0\} \text{ and } \mathcal{K}_+(\mathbf{X}) \ne \emptyset.$$

Definition. A positive integral on X is a non-zero linear functional $\mu : \mathcal{K}(\mathbf{X}) \to \mathbf{R}$ such that $\mu(f) \ge 0$ whenever $f \ge 0$. We remark in passing that both $\mathcal{K}(\mathbf{X})$ and $\mathbf{R}$ are partially ordered vector spaces, that is, vector spaces over $\mathbf{R}$ with a partial order $\le$ satisfying

(i) $f \leq g \Rightarrow f + h \leq g + h$,

(ii) $f \leq g \Rightarrow \lambda f \leq \lambda g$ for any real $\lambda \geq 0$.

The obvious morphisms to consider between such structures are the order-preserving linear functions. Thus a positive integral on X may be described as a non-trivial morphism of partially ordered vector spaces from $\mathcal{K}(X)$ to $\mathbf{R}$. Integrals do not, however, preserve the lattice operations.

Example. Let $X = \mathbf{R}$ and let a be any element of $\mathcal{K}_+(\mathbf{R})$. Then the functional $\mu : \mathcal{K}(\mathbf{R}) \to \mathbf{R}$ defined by

$$\mu(f) = \int_{-\infty}^{\infty} f(x)a(x)dx$$

is a positive integral on $\mathbf{R}$.

Definition. If G is a locally compact Hausdorff group, then a right Haar integral on G is a positive integral μ 'invariant under right translations'. This means that if $f, g \in \mathcal{K}(G)$ and if $f(x) = g(xa)$ for all $x \in G$, where a is a fixed element of G, then $\mu(f) = \mu(g)$.

Translation of functions introduces a new element of structure on $\mathcal{K}(G)$. For any function $f : G \to \mathbf{R}$, and any $a \in G$, we denote by f^a the function from G to $\mathbf{R}$ defined by $f^a(x) = f(xa^{-1})$. We call f^a the right translate of f by a. If f has support $S \subseteq G$ then f^a has support Sa, and it follows that if $f \in \mathcal{K}(G)$ then also $f^a \in \mathcal{K}(G)$. Thus the group G acts on the vector space $\mathcal{K}(G)$ making it a (right) G-module over $\mathbf{R}$, which means that

(i) $\mathcal{K}(G)$ is a vector space over $\mathbf{R}$;

(ii) for each $a \in G$, the map $f \mapsto f^a$ is a linear map from $\mathcal{K}(G)$ to itself;

(iii) $f^e = f$ for all $f \in \mathcal{K}(G)$;

(iv) $(f^a)^b = f^{ab}$ for all $f \in \mathcal{K}(G)$ and all $a, b \in G$.

For G-modules M, N, the natural definition of a morphism $M \to N$ is a linear map $\phi : M \to N$ such that $\phi(f^a) = (\phi(f))^a$ for all $f \in M$ and all $a \in G$. If we let G act trivially on $\mathbf{R}$ (i.e., $r^a = r$ for $r \in \mathbf{R}$, $a \in G$), then $\mathbf{R}$ is also a G-module and we may describe a Haar integral on G as an order-preserving morphism of G-modules from $\mathcal{K}(G)$ to $\mathbf{R}$.

2. Some results on approximation

Our aim is to prove that there exists an essentially unique right Haar integral on any locally compact group G. To do this we shall try to approximate functions $f \in \mathcal{K}(G)$ by sums of translates of scalar multiples of a function with small support. (Cf. approximation by step functions in the case of the Riemann integral.) The basis for such approximation is:

Proposition 19. Let $f, F \in \mathcal{K}_+(G)$. Then there exist real numbers $\alpha_1, \alpha_2, \ldots, \alpha_n \geq 0$ and $x_1, x_2, \ldots, x_n \in G$ such that

$$\sum_i \alpha_i F^{x_i} \geq f.$$

For any such α_i and x_i, and any right Haar integral μ on G,

$$\mu(f) \leq (\sum_i \alpha_i)\, \mu(F).$$

Proof. There exists $t \in G$ such that $F(t) \neq 0$. Let $\beta = \frac{1}{2} F(t) > 0$. Then there exists an open neighbourhood V of e in G such that $F(x) > \beta$ for all $x \in Vt$ (by the continuity of F). Let $C = \text{support}(f)$; then C is compact, so is covered by a finite number of sets Vc $(c \in C)$, say $C \subseteq Vc_1 \cup Vc_2 \cup \ldots \cup Vc_n$.

If $x \in Vc_i$ then $xc_i^{-1}t \in Vt$, so that $F(xc_i^{-1}t) > \beta$, i.e., $F^{x_i}(x) > \beta$, where $x_i = t^{-1}c_i$. But f is bounded (the set of its values in **R** is the image of a compact set!), say $f(x) \leq M$ for all $x \in G$. Hence

$$\sum_{i=1}^{n} \frac{M}{\beta} F^{x_i}(x) \geq M \geq f(x)$$

for all $x \in C$ (since $x \in$ some Vc_i), and $\sum \frac{M}{\beta} F^{x_i}(x) \geq f(x)$ trivially for $x \notin C$. This proves the first part of the proposition.

If now $f \leq \sum_i \alpha_i F^{x_i}$ we have

$$\begin{aligned} \mu(f) &\leq \mu(\sum_i \alpha_i F^{x_i}) \\ &= \sum_i \alpha_i \mu(F^{x_i}) \\ &= (\sum_i \alpha_i)\mu(F) \end{aligned}$$

for any right Haar integral μ.

Corollary. If $F \in \mathcal{K}_+(G)$ then $\mu(F) > 0$ for all right Haar integrals μ.

Proof. Since μ is non-trivial, there exists $f \in \mathcal{K}(G)$ such that $\mu(f) \neq 0$. Since either $\mu(f) > 0$ or $\mu(-f) > 0$, we have

$$\mu(|f|) \geq \max(\mu(f), \mu(-f)) > 0.$$

By the Proposition, there exist $\alpha_i \geq 0$ such that

$$0 < \mu(|f|) \leq (\sum \alpha_i)\mu(F).$$

So $\mu(F) > 0$.

For $f, F \in \mathcal{K}_+ = \mathcal{K}_+(G)$, let $(f : F)$ be defined as $(f : F) = \inf\{(\sum \alpha_i);\ \alpha_i > 0$, and for some $x_i \in G$, $\sum \alpha_i F^{x_i} \geq f\}$. By Proposition 19, $(f : F)$ exists and is non-negative; also if μ is any right Haar integral on G, then $\mu(f) \leq (f : F)\mu(F)$.

Proposition 20. Let $f, f_i, F, g \in \mathcal{K}_+(G)$. Then:

(i) $(f : F) > 0$,

(ii) for each $x \in G$, $(f^x : F) = (f : F)$,

(iii) $(f_1 + f_2 : F) \leq (f_1 : F) + (f_2 : F)$,

(iv) if $\alpha > 0$, $(\alpha f : F) = \alpha(f : F)$,

(v) $(f : F) \leq (f : g)(g : F)$.

Proof. (i): $f \leq \sum \alpha_i F^{x_i} \Rightarrow \sup(f) \leq (\sum \alpha_i)\sup(F)$
$\Rightarrow \sum \alpha_i \geq \sup(f)/\sup(F)$.

Hence $(f : F) \geq \sup(f)/\sup(F) > 0$.

(ii): $f \leq \sum \alpha_i F^{x_i} \Longleftrightarrow f^x \leq \sum \alpha_i F^{(x_i x)}$.

(iii): If $f_1 \leq A = \sum_i \alpha_i F^{x_i}$, and $f_2 \leq B = \sum_j \beta_j F^{y_j}$, then $f_1 + f_2 \leq A + B$, and $(f_1 + f_2 : F) \leq \sum_i \alpha_i + \sum_j \beta_j$. Thus, taking lower bounds, $(f_1 + f_2 : F) \leq (f_1 : F) + (f_2 : F)$.

(iv): $f \leq \sum \alpha_i F^{x_i}$ if and only if $\alpha f \leq \sum (\alpha\alpha_i) F^{x_i}$.

(v): Suppose that $f \leq \sum_i \alpha_i g^{x_i}$, $g \leq \sum_j \beta_i F^{y_j}$. Then $g^{x_i} \leq \sum_j \beta_j F^{y_j x_i}$, and $f \leq \sum_{i,j} \alpha_i \beta_j F^{(y_j x_i)}$. Thus $(f : F) \leq \sum_{i,j} \alpha_i \beta_j = (\sum_i \alpha_i)(\sum_j \beta_j)$, and hence $(f : F) \leq (f : g)(g : F)$.

We now normalize $(f : F)$ by fixing a <u>standard</u> function $g \in \mathcal{K}_+(G)$ and defining

$$\mu_F(f) = (f : F)/(g : F).$$

Proposition 20 implies immediately:

Proposition 20'. <u>Let</u> $f, f_i, F \in \mathcal{K}_+(G)$. <u>Then:</u>

(i) $\mu_F(f) > 0$,

(ii) <u>for</u> $x \in G$, $\mu_F(f^x) = \mu_F(f)$,

(iii) $\mu_F(f_1 + f_2) \leq \mu_F(f_1) + \mu_F(f_2)$,

(iv) <u>if</u> $\alpha > 0$, $\mu_F(\alpha f) = \alpha \mu_F(f)$,

(v) $(g : f)^{-1} \leq \mu_F(f) \leq (f : g)$.

Proposition 21. (Uniform continuity). <u>For any topological group</u> G, <u>and</u> $f \in \mathcal{K}(G)$, f <u>is uniformly continuous on</u> G, <u>that is, for arbitrary</u> $\varepsilon > 0$ <u>there is a symmetric open neighbourhood</u> V <u>of</u> e <u>in</u> G <u>such that</u>

$$|f(x) - f(y)| < \varepsilon$$

<u>whenever</u> $x, y \in G$ <u>and either</u> $xy^{-1} \in V$ <u>or</u> $x^{-1}y \in V$.

Proof. The function f has compact support C. For every $x \in C$ there is a neighbourhood $N(x)$ of e in G such that $|f(x) - f(y)| < \varepsilon/2$ whenever $y \in xN(x)$. Choose, for each $x \in C$, a symmetric open neighbourhood $M(x)$ of e such that $M(x)^2 \subseteq N(x)$. By compactness, there exist $x_1, x_2, \ldots, x_n \in C$ such that $x_1M(x_1) \cup x_2M(x_2) \cup \ldots \cup x_nM(x_n) \supseteq C$. Put $U = M(x_1) \cap M(x_2) \cap \ldots \cap M(x_n)$, and suppose that $x^{-1}y \in U$. If both x and y are outside C then $|f(x) - f(y)| = 0 < \varepsilon$. Otherwise we may suppose that $x \in C$, in which case $x \in x_iM(x_i)$ for some i. But then $y \in xU \subseteq x_iM(x_i)U \subseteq x_iN(x_i)$ and so

$$|f(x_i) - f(y)| < \varepsilon/2.$$

Since also $x \in x_i M(x_i) \subseteq x_i N(x_i)$, we have

$$|f(x_i) - f(x)| < \varepsilon/2,$$

whence

$$|f(x) - f(y)| < \varepsilon.$$

A similar argument shows that there is a symmetric neighbourhood U' of e such that

$$|f(x) - f(y)| < \varepsilon$$

whenever $xy^{-1} \in U'$. The neighbourhood $V = U \cap U'$ then has the required property.

Proposition 22. Let G be a locally compact Hausdorff topological group. Let $f_1, f_2 \in \mathcal{K}_+(G)$, and let $\varepsilon > 0$. Then there is an open neighbourhood V of e in G such that, whenever $F \in \mathcal{K}_+(G)$ has support $\subseteq V$, $\mu_F(f_1 + f_2) \geq \mu_F(f_1) + \mu_F(f_2) - \varepsilon$.

Proof. Let C be the union of the supports of f_1 and f_2. Then C is compact, and by Urysohn's Lemma we may select $q \in \mathcal{K}_+(G)$ such that $q(x) = 1$ for $x \in C$.

Let $\alpha > 0$, and define $p = f_1 + f_2 + \alpha q$, so that for $x \in C$, $p(x) \geq \alpha > 0$. Hence by continuity, $p(x) > 0$ on some open neighbourhood of C.

For $i = 1, 2$, define $h_i(x) = \begin{cases} f_i(x)/p(x) & \text{if } x \in C \\ 0 & \text{if } x \notin C. \end{cases}$ It is easy to see that h_i is continuous, and that support(h_i) is closed in the compact space C, hence is compact. So $h_i \in \mathcal{K}_+(G)$, and for all x,

$$0 \leq h_1(x) + h_2(x) \leq 1.$$

By Proposition 21, we can find a symmetric open neighbourhood V of e (depending on α) such that $|h_i(x) - h_i(y)| < \alpha/2$ $(i = 1, 2)$ whenever $xy^{-1} \in V$ or $x^{-1}y \in V$.

Suppose that $F \in \mathcal{K}_+(G)$, with support$(F) \subseteq V$, and that

$$p \leq \sum_j \beta_j F^{x_j} \quad (\text{with } \beta_j > 0,\ x_j \in G).$$

Then $F^{x_j}(t) = 0$ unless $t \in Vx_j$, and then $|h_i(x_j) - h_i(t)| < \alpha/2$; so

$$f_i = ph_i \le \sum_j \beta_j h_i F^{x_j}$$
$$\le \sum_j \beta_j (h_i(x_j) + \alpha/2) F^{x_j}.$$

It follows that $(f_i : F) \le \sum_j \beta_j (h_i(x_j) + \alpha/2)$, $(i = 1, 2)$. Thus $(f_1 : F) + (f_2 : F) \le (1 + \alpha) \sum \beta_j$, because $h_1(x_j) + h_2(x_j) \le 1$. Taking limits, $(f_1 : F) + (f_2 : F) \le (1 + \alpha)(p : F)$, and hence

$$\mu_F(f_1) + \mu_F(f_2) \le (1 + \alpha)\mu_F(p)$$
$$\le (1 + \alpha)(\mu_F(f_1 + f_2) + \alpha\mu_F(q))$$
$$= \mu_F(f_1 + f_2) + \alpha R,$$

where $R = \mu_F(f_1 + f_2) + (1 + \alpha)\mu_F(q)$
$$\le (f_1 + f_2 : g) + (1 + \alpha)(q : g).$$

But q, g, f_1, f_2 are fixed, so we may choose α small enough to ensure that $\alpha R \le \varepsilon$. The corresponding V is the neighbourhood of e required.

Corollary. Given $f_1, \ldots, f_n \in \mathcal{K}_+(G)$, and $\varepsilon > 0$, there is a symmetric open neighbourhood V of e such that, whenever $F \in \mathcal{K}_+(G)$ has support $\subseteq V$,

$$\mu_F(\sum f_i) \ge \sum \mu_F(f_i) - \varepsilon.$$

Proof. This is a straightforward induction on n.

3. Convex cones

The facts about the 'approximate Haar integrals' μ_F which we have now proved can be used in many ways to establish the existence of a Haar integral (see, e.g., Nachbin [7] or Hewitt and Ross [3]). The geometrical approach which we adopt here has the advantage of being more conceptual, though no shorter, than the standard methods. We first introduce some elementary notions from convexity theory and prove a general result on the existence of supporting hyperplanes of convex cones.

Definition. If V is a real vector space, a non-empty subset E of V is a <u>convex cone</u> if

(i) $\lambda > 0 \Rightarrow \lambda E \subseteq E$,

and (ii) $E + E \subseteq E$.

(It follows from (i) and (ii) that if λ, $\mu \geq 0$, and $\lambda + \mu = 1$, then for all $u, v \in E$, also $\lambda u + \mu v \in E$; i.e., E is convex in the usual sense.)

Example. $\mathcal{K}_+ = \mathcal{K}_+(G)$ is a convex cone in the real vector space $\mathcal{K} = \mathcal{K}(G)$.

The set E as above is an <u>open convex cone</u> if in addition

(iii) for any $c \in E$, and any $v \in V$, $\exists\, \delta > 0$ such that $c + \lambda v \in E$ whenever $|\lambda| < \delta$.

Note that if an open convex cone E contains 0, then $E = V$.

Because E is convex, (iii) is equivalent to

(iii)' every line (coset of a 1-dimensional subspace) in V meets E in an open interval (possibly empty or infinite).

Any non-zero linear functional μ on V has kernel a hyperplane (maximal proper subspace) H. Conversely, given a hyperplane H in V, there is a non-zero linear functional μ on V with kernel H, and such a μ is unique up to scalar multiple. (In fact $\mu : V \to V/H \cong \mathbf{R}$ is such a functional, and any μ' with kernel H factors uniquely through μ.)

Now take $V = \mathcal{K}$. If μ is a positive integral, then it maps $\mathcal{K}_+$ to the set of non-negative real numbers. If μ is a right Haar integral, then it maps $\mathcal{K}_+$ to the set of positive real numbers, and for $f \in \mathcal{K}$, $x \in G$, we have $\mu(f) = \mu(f^x)$. Hence $\mu(f - f^x) = 0$ and $f - f^x \in H = \mathrm{Ker}(\mu)$.

Let $\mathcal{L}$ be the $\mathbf{R}$-space (in $\mathcal{K}$) spanned by all functions $f - f^x$ (for $f \in \mathcal{K}$, $x \in G$). Then if μ is a right Haar integral with kernel H,

(i) $H \supseteq \mathcal{L}$,

(ii) $H \cap \mathcal{K}_+ = \emptyset$.

Conversely, suppose that H is a hyperplane in $\mathcal{K}$ satisfying (i) and (ii) above, and let μ be a linear functional with kernel H. Then either μ or $-\mu$ is a right Haar integral. For if $f \in \mathcal{K}$ and $x \in G$, then $f - f^x \in H$, so $\mu(f) = \mu(f^x)$, that is, μ is right-translation-invariant. Also, given $f_1, f_2 \in \mathcal{K}_+$, and $0 \leq \lambda \leq 1$, we have $\lambda f_1 + (1 - \lambda) f_2 \in \mathcal{K}_+$. Thus, by condition (ii), $\lambda\mu(f_1) + (1 - \lambda)\mu(f_2) \neq 0$ for all $0 \leq \lambda \leq 1$. It follows that $\mu(f_1)$ and $\mu(f_2)$ have the same sign, and therefore either μ or $-\mu$ maps $\mathcal{K}_+$ to the positive reals, hence is a Haar integral.

Proposition 23. _Let_ V _be a real vector space,_ E _a convex cone in_ V, _and_ W _a subspace of_ V _which does not meet_ E. _Then:_

(i) $D = E + W$ _is a convex cone which does not meet_ W.

(ii) _If_ E _is open, so is_ D.

(iii) _If_ E _is open, there is a hyperplane_ $H \supseteq W$ _such that_ $E \cap H = \emptyset$.

(iv) _If_ E _is open, and its complement_ $V \backslash E$ _is convex, there is a unique_ H _as in (iii)._

Proof. (i) $\lambda > 0 \Rightarrow \lambda D \subseteq \lambda E + \lambda W \subseteq D$, and $D + D \subseteq E + E + W + W \subseteq D$.

(ii) Let $d = e + w \in D$, where $e \in E$, $w \in W$. Let $v \in V$, and choose $\delta > 0$ such that $e + \lambda v \in E$ whenever $|\lambda| < \delta$. Then $d + \lambda v \in E + W = D$ for such λ.

(iii) The set S of subspaces $U \supseteq W$ such that $U \cap E = \emptyset$ is non-empty and inductively ordered by inclusion (note that the union of a chain of subspaces is also a subspace). Hence, by Zorn's lemma, there is a maximal element $H \in S$. We have to show that H is a hyperplane.

Maximality of H says that if $v \notin H$, then $H + \mathbf{R}v$ meets E, hence $\mathbf{R}v$ meets $D = H + E$. By (i) and (ii) D is an open convex cone not meeting H; hence $0 \notin D$, and so either $v \in D$ or $-v \in D$.

Thus $V = D \cup H \cup -D$, and this union is disjoint. (For $D \cap H = \emptyset \Rightarrow (-D) \cap H = \emptyset$; and if $u \in D \cap (-D)$, then $u \in D$ and $-u \in D$, so $0 = u + (-u) \in D$, a contradiction.)

To show that H is a hyperplane, we need to prove that $v \notin H \Rightarrow H + \mathbf{R}v = V$. We therefore suppose that $v \notin H$ and $w \notin H + \mathbf{R}v$. Then, because $w \notin H$ we have $w \in D$ or $w \in -D$, and similarly for v. We may assume that $v \in D$, $w \in -D$. Since D and $-D$ are open convex cones, the line joining v and w meets these cones in two disjoint, non-empty open intervals. The union of these intervals cannot exhaust the line, so some point of the segment between v and w, say $\lambda v + \mu w$ $(\lambda + \mu = 1)$, must lie in H. But $v \notin H$, so $\mu \neq 0$ and hence $w \in H + \mathbf{R}v$, a contradiction.

(iv) E is an open convex cone, and $E^* = V\backslash E$ is a convex cone, whence so are $(-E^*)$, and $V\backslash(E \cup -E) = E^* \cap (-E^*) = H$, say. Clearly $u \in H \Rightarrow -u \in H$, and H is a subspace. Also $V = E \cup H \cup (-E)$, a disjoint union, and so proceeding as in (iii), we see that H is a hyperplane. Further, $W \cap E = \emptyset$, so $W \cap (-E) = \emptyset$, and $W \subseteq H$. Similarly if H' is any subspace containing W and not meeting E, then $H' \subseteq H$. In particular if H' is also a hyperplane, then $H' = H$, so H is unique.

4. The existence of Haar integrals

Theorem 5. Let G be a locally compact Hausdorff topological group. Then there is a right Haar integral μ on G. (Similarly there is a left Haar integral ν, but note that $\mu \neq \nu$ in general.)

Proof. Let $\mathcal{K} = \mathcal{K}(G)$, $\mathcal{K}_+ = \mathcal{K}_+(G)$, and let $\mathcal{L}$ be the **R**-subspace of $\mathcal{K}$ spanned by all functions $f - f^x$ $(f \in \mathcal{K},\ x \in G)$.

Every $f \in \mathcal{K}$ can be written as $f_1 - f_2$, with $f_1, f_2 \in \mathcal{K}_+$ (e.g. $f_1 = \frac{1}{2}(|f| + f)$, $f_2 = \frac{1}{2}(|f| - f)$). Also observe that if $f \in \mathcal{K}_+$ and $x \in G$, then $f^x = h \in \mathcal{K}_+$, and $f^x - f = h - h^{x^{-1}}$. Thus every $f \in \mathcal{L}$ has the form $f = \sum_i (f_i - f_i^{x_i})$, for some $f_i \in \mathcal{K}_+$, and $x_i \in G$.

Lemma A. $\mathcal{K}_+ \cap \mathcal{L} = \emptyset$.

Proof. This is essentially Proposition 22, Corollary. Suppose that f, $f_i \in \mathcal{K}_+$, $x_i \in G$, and

$$f = \sum_i (f_i - f_i^{x_i}).$$

Then $f + \sum_i f_i^{x_i} = \sum_i f_i$, and we are looking for a contradiction.

Let $\varepsilon > 0$. By the corollary to Proposition 22 we can find an open set V such that, whenever $F \in \mathcal{K}_+$ has support contained in V,

$$\mu_F(f + \sum_i f_i^{x_i}) \geq \mu_F(f) + \sum_i \mu_F(f_i^{x_i}) - \varepsilon.$$

Such an F exists. For as G is locally compact, we can find an open set U with compact closure $\overline{U} \subseteq V$. By Urysohn's Lemma, we can find

$F \in \mathcal{K}_+$ such that support$(F) \subseteq V$.

With such an F, we have

$$\begin{aligned}\sum_i \mu_F(f_i) &\geq \mu_F(\sum_i f_i) \\ &= \mu_F(f + \sum_i f_i^{x_i}) \\ &\geq \mu_F(f) + \sum_i \mu_F(f_i^{x_i}) - \varepsilon \\ &\geq (g : f)^{-1} + \sum_i \mu_F(f_i) - \varepsilon, \text{ by Proposition 20'.}\end{aligned}$$

Hence $\varepsilon \geq (g : f)^{-1}$ for all $\varepsilon > 0$, a contradiction.

Lemma B. $\mathcal{C} = \mathcal{K}_+ + \mathcal{L}$ is open.

Proof. This is essentially Proposition 19. We know that $\mathcal{C}$ is a convex cone, and $\mathcal{C} \cap \mathcal{L} = \emptyset$, by Proposition 23. Let $f \in \mathcal{C}$, say $f = p + q$, $p \in \mathcal{K}_+$ and $q \in \mathcal{L}$. Let $k \in \mathcal{K}$. By Proposition 19 we can find $\alpha_i > 0$, $x_i \in G$, such that $|k| \leq \sum \alpha_i p^{x_i}$. Thus for $\lambda > 0$,

$$\begin{aligned}f \pm \lambda k &\geq p + q - \lambda \sum \alpha_i p^{x_i} \\ &= p(1 - \lambda \sum \alpha_i) + q',\end{aligned}$$

where $q' = q + \lambda \sum \alpha_i (p - p^{x_i}) \in \mathcal{L}$.

Hence $f \pm \lambda k \in \mathcal{C}$ whenever $1 - \lambda \sum \alpha_i > 0$, i.e., whenever $0 < \lambda < (\sum \alpha_i)^{-1}$, which shows that $\mathcal{C}$ is open.

It now follows by Proposition 23 that there is a hyperplane H in $\mathcal{K}$ such that $H \supseteq \mathcal{L}$ and $H \cap \mathcal{C} = \emptyset$ (so in particular $H \cap \mathcal{K}_+ = \emptyset$). By the remarks preceding Proposition 23, H determines a right Haar integral on G.

5. The uniqueness of Haar integrals

We shall use Theorem 5 to prove the uniqueness of the right Haar integral. It is possible, at the expense of some extra detailed analysis, to prove both existence and uniqueness at the same time, but the effort seems hardly worth while. The only advantage is that it does away with the use of the choice axiom in the proof of existence; but since Urysohn's

Lemma is essential and relies on the choice axiom this is not a genuine logical advantage.

Proposition 24 (Dieudonné). Let $f \in \mathcal{K}_+(G)$, where G is a locally compact Hausdorff group, and let $C = \text{support}(f)$; let W be any neighbourhood of e in G. Then there exist $x_1, x_2, \ldots, x_n \in C$ and functions $f_1, f_2, \ldots, f_n \in \mathcal{K}_+(G)$ such that

(i) $f = \sum f_i$

(ii) $\text{support}(f_i) \subseteq Wx_i$ for each i.

Proof. By Proposition 17, there exists a compact neighbourhood N of e in G contained in the interior of W. Since C is compact there exist $x_1, \ldots, x_n \in C$ such that $Nx_1, \ldots, Nx_n$ cover C. Choose $h_i \in \mathcal{K}_+(G)$ such that $h_i(x) = 1$ for $x \in Nx_i$ and $\text{support}(h_i) \subseteq Wx_i$. (Urysohn's Lemma guarantees the existence of the h_i.) Put $h = \sum h_i$; then $h(x) \geq 1$ for all $x \in C$ and we may define

$$f_i(x) = \begin{cases} f(x)h_i(x)/h(x) & \text{if } x \in C \\ 0 & \text{if } x \notin C. \end{cases}$$

Then f_i is continuous, $\text{support}(f_i) \subseteq \text{support}(h_i) \subseteq Wx_i$ and $\sum f_i = f$.

Proposition 25 (Uniform approximation theorem). Let $f \in \mathcal{K}_+(G)$ and $\varepsilon > 0$. Then there exists a neighbourhood V of e in G such that for every symmetric $F \in \mathcal{K}_+(G)$ (i.e., $F(x^{-1}) = F(x)$) with $\text{support}(F) \subseteq V$ there exist real numbers $\alpha_1, \ldots, \alpha_n \geq 0$ and $x_1, \ldots, x_n \in G$ such that

$$\left| f(x) - \sum \alpha_i F^{x_i}(x) \right| \leq \varepsilon$$

for all $x \in G$.

Proof. Choose a neighbourhood V of e so that

$$|f(x) - f(y)| < \frac{\varepsilon}{2} \text{ whenever } y \in Vx$$

(by Proposition 21), and let $F \in \mathcal{K}_+(G)$ be symmetric with $\text{support}(F) \subseteq V$.

Then we have $\text{support}(F^x) \subseteq Vx$, so

$$|f(x) - f(y)| F^x(y) \leq \frac{\varepsilon}{2} F^x(y)$$

for all $x, y \in G$. Viewing the two sides of this inequality as functions of y, we may rewrite it as

$$|f(x)F^x - fF^x| \leq \frac{\varepsilon}{2} F^x \quad \text{for all } x \in G. \tag{1}$$

Now let $\delta > 0$ and choose another neighbourhood W of e such that

$$|F(y) - F(z)| < \delta \quad \text{for all } y \in Wz.$$

Then

$$|F^x(y) - F^x(z)| < \delta \quad \text{for all } y \in Wz \text{ and all } x \in G.$$

By Proposition 24 (applied to this W) we can find $x_1, \ldots, x_n \in G$, $f_1, \ldots, f_n \in \mathcal{K}_+(G)$ such that $f = \sum f_i$ and $\text{support}(f_i) \subseteq Wx_i$. For each i it follows that

$$f_i(y) |F^x(y) - F^x(x_i)| \leq \delta f_i(y) \quad \text{for all } x, y \in G.$$

Now we observe that, since F is symmetric, $F^{x_i}(x) = F^x(x_i)$, and summing over i we get

$$|f(y)F^x(y) - \sum f_i(y)F^{x_i}(x)| \leq \delta f(y)$$

for all $x, y \in G$, from which follows

$$|fF^x - \sum f_i F^{x_i}(x)| \leq \delta f \quad \text{for all } x \in G. \tag{2}$$

Combining (1) and (2) we get

$$|f(x)F^x - \sum f_i F^{x_i}(x)| \leq \frac{\varepsilon}{2} F^x + \delta f \quad \text{for all } x \in G. \tag{3}$$

Now, by Theorem 5, there exists a right Haar integral μ on G. We apply μ to (3) to obtain

$$|f(x)\mu(F) - \sum \mu(f_i)F^{x_i}(x)| \leq \frac{\varepsilon}{2} \mu(F) + \delta\mu(f)$$

$$\leq \frac{\varepsilon}{2}\,\mu(F) + \delta(f : F)\mu(F)$$

for all $x \in G$. Now divide by $\mu(F)$ and put $\alpha_i = \mu(f_i)/\mu(F)$. This gives

$$|f(x) - \sum \alpha_i F^{x_i}(x)| \leq \frac{\varepsilon}{2} + \delta(f : F) \text{ for all } x \in G.$$

But δ is independent of F, so we can choose δ small enough for $\delta(f : F) < \frac{\varepsilon}{2}$. The resulting α_i and x_i give the required approximation.

Corollary. Let $f \in \mathcal{K}(G)$ and let $\mathcal{C}$ be the cone $\mathcal{K}_+(G) + \mathcal{L}$ (as in the proof of Theorem 5). Then there exists $h \in \mathcal{K}_+(G)$ such that for every $\varepsilon > 0$ either $f + \varepsilon h \in \mathcal{C}$ or $f - \varepsilon h \in -\mathcal{C}$.

Proof. Let $C = \text{support}(f)$ and let D be a compact neighbourhood of C (Proposition 17). Let $h \in \mathcal{K}_+(G)$ with $h(x) > 2$ for $x \in D$ (such an h exists by Urysohn's lemma). Write f in the form $f_1 - f_2$, $f_1, f_2 \in \mathcal{K}_+(G)$, and apply the approximation theorem to f_1 and f_2, choosing the respective neighbourhoods V_1, V_2 of e small enough for $V_1C \subset D$, $V_2C \subset D$ (this is possible by Proposition 11). Let $V \subseteq V_1 \cap V_2$ be symmetric and choose $F_0 \in \mathcal{K}_+(G)$ with $\text{support}(F_0) \subseteq V$ (by Urysohn). Put $F(x) = F_0(x) + F_0(x^{-1})$. Then $F \in \mathcal{K}_+(G)$, F is symmetric, and $\text{support}(F) \subseteq V$. Both f_1 and f_2 can be approximated as in the Proposition, using this F. Hence there exist $x_1, \ldots, x_n \in C$ and $\alpha_1, \alpha_2, \ldots, \alpha_n \in \mathbf{R}$ (not now necessarily ≥ 0) such that

$$|f(x) - \sum \alpha_i F^{x_i}(x)| \leq 2\varepsilon \quad \text{for all } x \in G.$$

Since f and F^{x_i} vanish outside D it follows that

$$|f - \sum \alpha_i F^{x_i}| < \varepsilon h.$$

Put $\alpha = \sum \alpha_i$; then $k = \alpha F - \sum \alpha_i F^{x_i} = \sum \alpha_i (F - F^{x_i}) \in \mathcal{L}$, and

$$f - \varepsilon h < \alpha F - k < f + \varepsilon h.$$

Now either $\alpha \geq 0$, in which case

$$f + \varepsilon h > -k \in \mathcal{L},$$

so that

$$f + \varepsilon h \in \mathcal{L} + \mathcal{K}_+(G) = \mathcal{C},$$

or $\alpha \le 0$, in which case

$$f - \varepsilon h < -k \in \mathcal{L}$$

and

$$f - \varepsilon h \in -\mathcal{C}.$$

Theorem 6. _If_ μ, μ' _are right Haar integrals on a locally compact Hausdorff group_ G _then there exists_ $\lambda > 0$ _such that_ $\mu' = \lambda\mu$.

(Remark: The equivalent statement holds, of course, for left Haar integrals. But the right and left Haar integrals of a group need not be the same.)

Proof. By Proposition 23(iv) it is enough to show that the complement of $\mathcal{C}$ is convex, since we would then obtain a unique hyperplane H containing $\mathcal{L}$ and not meeting $\mathcal{C}$; since $\ker \mu$ and $\ker \mu'$ are such hyperplanes, they must coincide; therefore $\mu = \lambda\mu'$ and λ will obviously be positive.

In other words, it is enough to show that if $f_1, f_2 \in \mathcal{K}(G)$ and $f_1 + f_2 \in \mathcal{C}$ then $f_1 \in \mathcal{C}$ or $f_2 \in \mathcal{C}$. Suppose that $f_1 + f_2 \in \mathcal{C}$ and let $f_1 - f_2 = f$. By Corollary to Proposition 25 there exists $h \in \mathcal{K}_+(G)$ such that for all $\varepsilon > 0$ either $f + \varepsilon h \in \mathcal{C}$ or $f - \varepsilon h \in -\mathcal{C}$, i.e., either

$$f_1 - f_2 + \varepsilon h \in \mathcal{C} \text{ or } f_2 - f_1 + \varepsilon h \in \mathcal{C}. \qquad (*)$$

But $\mathcal{C}$ is open; so there exists $\varepsilon > 0$ such that $f_1 + f_2 - \varepsilon h \in \mathcal{C}$; adding this to (*) we get either $2f_1 \in \mathcal{C}$ or $2f_2 \in \mathcal{C}$, i.e., $f_1 \in \mathcal{C}$ or $f_2 \in \mathcal{C}$.

6. Integrals on product groups

We now exhibit the relationship between the right Haar integrals μ on G, ν on H and the right Haar integrals on $G \times H$.

Notation. We shall write the more explicit form

$$\int_G f(x)d\mu(x)$$

for what we have so far denoted by $\mu(f)$.

Proposition 26 (Fubini's theorem). Let G, H be locally compact Hausdorff groups with right Haar integrals μ, ν respectively. Let $f \in \mathcal{K}(G \times H)$. Then the integrals

$$\rho(f) = \int_G \int_H f(x, y)d\nu(y)d\mu(x)$$

and

$$\sigma(f) = \int_H \int_G f(x, y)d\mu(x)d\nu(y)$$

exist, are equal, and are right Haar integrals on $G \times H$.

Proof. There exists a compact set $C \subseteq H$ such that $f(x, y) = 0$ for $y \notin C$ (project support(f) onto H). So, for each $x \in G$, the function $\phi_x : y \mapsto f(x, y)$ is in $\mathcal{K}(H)$, and $\psi(x) = \nu(\phi_x)$ exists. To show that $\rho(f)$ exists, we must show that $\psi \in \mathcal{K}(G)$. Clearly ψ has compact support (again by projection!). Given $\varepsilon > 0$ there exists a neighbourhood V of e in G such that

$$|f(x', y) - f(x, y)| < \varepsilon \text{ for all } x' \in Vx, \ y \in H,$$

since f is uniformly continuous on $G \times H$. Choose $k \in \mathcal{K}_+(H)$ such that $k(y) = 1$ for $y \in C$. Then

$$|f(x', y) - f(x, y)| \leq \varepsilon k(y) \text{ for all } y \in H, \ x' \in Vx.$$

Apply ν to get

$$|\psi(x') - \psi(x)| \leq \varepsilon\nu(k) \quad \text{for all } x' \in Vx,$$

which is just the statement of continuity for ψ.

So $\psi \in \mathcal{K}(G)$ and $\rho(f) = \mu(\psi)$ exists; this holds for any $f \in \mathcal{K}(G \times H)$. Now ρ is clearly linear, positive and right-invariant; therefore it is a right Haar integral on $G \times H$.

Similarly σ is a right Haar integral on $G \times H$. So for some $\lambda > 0$

$$\sigma(f) = \lambda\rho(f) \quad \text{for all } f \in \mathcal{K}(G \times H).$$

To show that $\lambda = 1$, choose $p \in \mathcal{K}_+(G)$, $q \in \mathcal{K}_+(H)$ and put

$$f(x, y) = p(x)q(y) \quad \text{for all } x \in G, \ y \in H.$$

Then, for this f,

$$\rho(f) = \sigma(f) = \mu(p)\nu(q) \neq 0.$$

It follows that $\lambda = 1$, and hence that $\rho = \sigma$.

7. Unimodular groups

Let G be a locally compact Hausdorff group, and let μ be a right Haar integral on G. Let $s \in G$ and consider the left translate ${}^s f$ of f by s (i.e. ${}^s f(x) = f(s^{-1}x)$). In general $\mu({}^s f) \neq \mu(f)$ (example later!). However, we can define a new linear functional ${}^s\mu$ on $\mathcal{K}(G)$ by ${}^s\mu(f) = \mu({}^s f)$. This functional ${}^s\mu$ is, in fact, a positive integral (check!) and is right invariant:

$${}^s\mu(f^t) = \mu({}^s(f^t)) = \mu(({}^s f)^t) = \mu({}^s f) = {}^s\mu(f).$$

Thus ${}^s\mu$ is a right Haar integral and so there exists a positive real number $\Delta(s)$ such that ${}^s\mu = \Delta(s)\mu$; $\Delta(s)$ is independent of the particular right Haar integral μ chosen. (Exercise.)

The function $\Delta : G \to \mathbf{R}^{pos}$ so defined satisfies

$$\Delta(st) = \Delta(s)\Delta(t) = \Delta(ts)$$

and is continuous (Exercise). It is, therefore, a morphism in $\mathcal{TG}$. Δ is called the right modular function of G. A group with $\Delta(s) = 1$ for all s is called unimodular.

If G is unimodular then ${}^s\mu = \mu$ for all s, so μ is left invariant; thus the right and left Haar integrals coincide; conversely, if the left and right Haar integrals on G are the same, then G is unimodular. Obviously any Abelian group is unimodular. A little unexpectedly we also have:

Proposition 27. All compact Hausdorff groups are unimodular.

First proof. Let $f(x) = 1$ for all $x \in G$; then $f \in \mathcal{K}(G)$ and is invariant under all translations. Hence

$$\Delta(s)\mu(f) = \mu({}^{s}f) = \mu(f) \neq 0, \quad \text{so} \quad \Delta(s) = 1 \ \text{for all} \ s \in G.$$

Second proof. $\Delta : G \to \mathbf{R}^{pos}$ is a morphism in $\mathcal{TG}$; its image must be a compact subgroup of $\mathbf{R}^{pos}$. But the only compact subgroup of $\mathbf{R}^{pos}$ is $\{1\}$ (Exercise).

Note: For a compact group G we can always normalize the Haar integral so that

$$\int_G 1 \, d\mu = 1.$$

The relation between right and left Haar integrals on a group G can be made explicit via the modular function. For $f \in \mathcal{K}(G)$ let $\hat{f}$ be defined by $\hat{f}(x) = f(x^{-1})$; then $\hat{f} \in \mathcal{K}(G)$. Define ν by $\nu(f) = \mu(\hat{f})$, where μ is a right Haar integral. Immediately, ν is a positive integral. It is also left-invariant:

$$\widehat{{}^{s}f}(x) = {}^{s}f(x^{-1}) = f(s^{-1}x^{-1}) = \hat{f}(xs) = \hat{f}^{s^{-1}}(x),$$

whence

$$\nu({}^{s}f) = \mu(\widehat{{}^{s}f}) = \mu(\hat{f}^{s^{-1}}) = \mu(\hat{f}) = \nu(f).$$

Therefore ν is a left Haar integral.

An immediate consequence is that μ is invariant under inversion (i.e., $\mu(\hat{f}) = \mu(f)$ for all f) if and only if the group is unimodular.

We get the left modular function as follows: by an argument similar to the one above, we see that $\widehat{f^{s}} = {}^{s^{-1}}\hat{f}$, whence

$$\nu(f^{s}) = \mu(\widehat{f^{s}}) = \mu({}^{s^{-1}}\hat{f}) = \Delta(s^{-1})\mu(\hat{f}) = \Delta(s^{-1})\nu(f).$$

Thus the left modular function is $\hat{\Delta}$; $\hat{\Delta}(s) = \Delta(s^{-1}) = \Delta(s)^{-1}$.

Proposition 28. For a right Haar integral μ with right modular function Δ,

$$\int_G \hat{f}(x) d\mu(x) = \int_G \frac{f(x)}{\Delta(x)} d\mu(x),$$

and this formula defines a left Haar integral, $\int_G f(x) d\nu(x)$.

Proof. The left hand side is $\mu(\hat{f}) = \nu(f)$ and we have seen that ν is a left Haar integral. Let $f^*(x) = f(x)/\Delta(x)$; then the right hand side is $\mu(f^*) = \rho(f)$, say. Now ρ is clearly a positive integral. It is also left invariant:

$$({}^s f)^*(x) = \frac{{}^s f(x)}{\Delta(x)} = \frac{f(s^{-1}x)}{\Delta(x)} = \frac{1}{\Delta(s)} \frac{f(s^{-1}x)}{\Delta(s^{-1}x)} = \frac{1}{\Delta(s)} {}^s(f^*)(x),$$

whence

$$\rho({}^s f) = \mu(({}^s f)^*) = \mu\left(\frac{1}{\Delta(s)} {}^s(f^*)\right) = \frac{1}{\Delta(s)} \mu({}^s(f^*)) = \mu(f^*) = \rho(f).$$

So, by Theorem 6, there exists $\lambda > 0$ such that

$$\rho(f) = \lambda\nu(f) \text{ for all } f \in \mathcal{K}(G).$$

To show that $\lambda = 1$ we must choose an appropriate function and compute the two sides. Let $\varepsilon > 0$ and choose a neighbourhood V of e in G such that

$$\left|1 - \frac{1}{\Delta(s)}\right| < \varepsilon \text{ whenever } s \in V$$

(this can be done since Δ is continuous). There exists a symmetric function $f \in \mathcal{K}_+(G)$ with $\text{support}(f) \subseteq V$ and $\mu(f) = 1$. For this f,

$$|\hat{f} - f^*| = \left|\left(1 - \frac{1}{\Delta}\right)f\right| < \varepsilon f.$$

We can now apply μ to obtain

$$|\nu(f) - \rho(f)| \leq \varepsilon\mu(f) = \varepsilon,$$

and since

$$\nu(f) = \mu(\hat{f}) = \mu(f) = 1 \text{ and } \rho(f) = \lambda\nu(f) = \lambda,$$

we have

$$|1 - \lambda| < \varepsilon \text{ for any } \varepsilon > 0,$$

i. e. $\lambda = 1$. This proves the proposition.

IV · Examples and applications

In this final chapter we aim to give the reader some idea of what the Haar integral looks like in various special cases and how it is used in practice. The material is informally presented so as to convey information more rapidly than would be possible in a self-contained, rigorous account. It will be an advantage for the reader to have some knowledge of such topics as Lebesgue integration, measure theory, normed linear spaces and the representation theory of finite groups. However, this is not essential for a general understanding of the chapter, and it is hoped that the items chosen will act as a stimulus to further reading.

1. Extensions of Haar integrals

Let G be a locally compact, Hausdorff group and let μ be a right Haar integral on G. Then $\mathcal{K} = \mathcal{K}(G)$ is a normed space with respect to $\|f\| = \mu(|f|)$; however, $\mathcal{K}$ is not complete. There is a standard procedure, due to Daniell, for completing $\mathcal{K}$ which we now describe briefly.

Let U be the set of all functions $f : G \to \mathbf{R} \cup \{+\infty\}$ which are pointwise limits of increasing sequences $\{f_n\}$, $f_n \in \mathcal{K}$. For $f \in U$, define $\mu(f) = \lim_{n\to\infty} \mu(f_n)$ (possibly $+\infty$). The value of $\mu(f)$ is independent of the sequence chosen. Let $-U = \{-f;\ f \in U\}$, the set of pointwise limits of decreasing sequences of functions in $\mathcal{K}$. For $f \in -U$, define $\mu(f) = -\mu(-f)$. This definition is consistent with the previous one.

A function $f : G \to \mathbf{R} \cup \{\pm\infty\}$ is said to be <u>summable</u> if

(i) there exist functions $g \in -U$ and $h \in U$ such that

$$g \le f \le h,$$

and (ii) $\sup_g \mu(g) = \inf_h \mu(h) < \infty$, where the inf and sup are taken over all g and h in (i).

If f is summable we define $\mu(f) = \sup_g \mu(g) = \inf_h \mu(h)$. Denote by $\mathcal{L}^1 = \mathcal{L}^1(G)$ the set of all summable functions. If $f \in \mathcal{L}^1$ then also $|f| \in \mathcal{L}^1$. For any $f \in \mathcal{L}^1$, define $\|f\| = \mu(|f|)$. Then $\mathcal{L}^1$ is a semi-normed space (there may exist non-zero functions whose norm is 0), and is complete (Riesz-Fischer Theorem).

A null-function in $\mathcal{L}^1$ is a function f such that $\|f\| = 0$. If we now define $L^1 = \mathcal{L}^1/\{\text{null-functions}\}$, then L^1 is a Banach space. Similarly, we define $\mathcal{L}^p$ (for $p \geq 1$) to be the space of all functions f such that $|f|^p$ is summable and define $\|f\|_p = \{\mu(|f|^p)\}^{1/p}$. Then $L^p = \mathcal{L}^p/\{\text{null-functions}\}$ is again a Banach space, and $\mathcal{K}$ is a dense subset of L^p, for each p. Further, $L^2 = L^2(G)$ is a Hilbert space, with $\langle f, g\rangle = \int_G f(x)g(x)d\mu(x)$. This theory extends easily to complex-valued functions, giving complex Banach spaces.

We can now define a right Haar measure on G, that is, a measure invariant under right translations. We say that $A \subseteq G$ is measurable if χ_A, the characteristic function of A, is summable. The measure of A is then defined to be $\int_G \chi_A(x)d\mu(x)$. It is possible to recover the Haar integral from the Haar measure by the standard procedure for the Lebesgue integral. For further details see Nachbin [7] and Widom [12].

2. Examples of Haar integrals

1. $G = (\mathbf{R}^n, +)$. This group is unimodular (being Abelian) and the extended Haar integral is $\int_{\mathbf{R}^n} f(x)dx$, the n-dimensional Lebesgue integral.

2. $G = S^1 = \mathbf{R}/\mathbf{Z}$. Again this is a unimodular group. A function $f : S^1 \to \mathbf{R}$ is essentially the same thing as a periodic function $f^* : \mathbf{R} \to \mathbf{R}$ with period 1. The Haar integral μ is given by

$$\mu(f) = \int_0^1 f^*(x)dx.$$

Alternatively, we may take $S^1 = \{e^{i\theta};\ 0 \leq \theta \leq 2\pi\}$ and the Haar integral is then $\mu(f) = \frac{1}{2\pi}\int_0^{2\pi} f(e^{i\theta})d\theta$, where the factor $\frac{1}{2\pi}$ produces the normalized integral (S^1 being compact).

3. Induced integrals. (i) Let μ be a right Haar integral on the locally compact, Hausdorff group G and let H be an open subgroup

of G. Any $f \in \mathcal{K}(H)$ extends to $\bar{f} \in \mathcal{K}(G)$ by the definition

$$\bar{f}(x) = \begin{cases} f(x) & \text{if } x \in H \\ 0 & \text{if } x \notin H. \end{cases}$$

(Exercise: check this!)

Defining $\mu'(f) = \mu(\bar{f})$ we get a right-invariant integral and hence a right Haar integral on H.

<u>N. B.</u> This method does not work for all subgroups or even for all closed subgroups; for example the Lebesgue integral on $\mathbf{R}$ induces on $\mathbf{Z}$ by this method the trivial integral, but the Haar integral is never trivial.

(ii) Suppose that G is a locally compact, Hausdorff group and N is a closed, normal subgroup of G. Then $Q = G/N$ is again locally compact and Hausdorff. Suppose that μ_N and μ_Q are right Haar integrals on N and Q respectively. We can construct μ_G, a right Haar integral on G, as follows.

Let $f \in \mathcal{K}(G)$, and for $x \in G$ define

$$f^*(x) = \int_N f(yx)d\mu_N(y).$$

Because μ_N is right-invariant, $f^*(x)$ is constant on each coset of N. Hence f^* induces a function $\bar{f}: Q \to \mathbf{R}$ defined by $\bar{f}(Nx) = f^*(x)$. It follows that $\bar{f} \in \mathcal{K}(Q)$ (check!).

Defining $\mu_G(f) = \mu_G(\bar{f})$, it is obvious that μ_G is a positive integral and easy to verify that μ_G is right-invariant. Thus it is a right Haar integral. (A particular case is the Haar integral on a product $A \times B$.)

4. <u>Finite discrete groups.</u> For a finite discrete group G, $\mathcal{K}(G) = L^1(G)$ is the set of <u>all</u> functions $f : G \to \mathbf{R}$. The normalized (left and right) Haar integral is given by

$$\mu(f) = \frac{1}{n} \sum_{x \in G} f(x),$$

where n is the order of G.

Now $L^1(G)$ is isomorphic to the group ring $\mathbf{R}G$ (which is a real vector space with basis G), so the multiplication in $\mathbf{R}G$, induced by the multiplication in G, gives a multiplication on $L^1(G)$ as follows. If

$f, g \in L^1(G)$ then the corresponding elements of $\mathbf{R}G$ are $\sum f(x).x$ and $\sum g(x).x$. Define $f * g \in L^1(G)$ to be the function corresponding to their product:

$$\sum_{x \in G} \sum_{y \in G} f(x)\, g(y).\, xy = \sum_{z \in G} \{ \sum_{x \in G} f(x) g(x^{-1}z) \} z$$

$$= \sum_{z \in G} \{ \sum_{y \in G} f(zy^{-1}) g(y) \} z,$$

i.e., $f * g = h$ where $h(z) = \sum_{x \in G} f(x) g(x^{-1}z) = \sum_{y \in G} f(zy^{-1}) g(y)$.

By analogy, for any locally compact Hausdorff group G, with right Haar integral μ, we can define a multiplication $*_\mu$ on $L^1(G)$ by

$$(f *_\mu g)(z) = \int_G f(zy^{-1}) g(y) d\mu(y).$$

(Since this definition is for f, g in $\mathcal{L}^1$ one needs to show that it behaves correctly on taking quotients with respect to the null functions.) Similarly, for the left Haar integral ν, define:

$$(f *_\nu g)(z) = \int_G f(x) g(x^{-1}z) d\nu(x).$$

Exercises. (i) $(f *_\mu g)(z) = \int_G f(y^{-1}) g(yz) d\mu(y)$.

(ii) $*_\mu$ is associative on L^1, and $\|f *_\mu g\|_1 \leq \|f\|_1 \cdot \|g\|_1$. This shows that L^1 is a Banach algebra.

(iii) L^1 is commutative if and only if G is Abelian.

(iv) L^1 has an identity element if and only if G is discrete.

5. Arbitrary discrete groups. For any discrete group G, $\mathcal{K}(G)$ is the set of all functions with finite support, and the Haar integral is $\mu(f) = \sum_{x \in G} f(x)$. This is invariant under inversion, so G is unimodular.

$L^1(G)$ is the set of all functions f with countable support S, such that

$$\sum_{x \in S} |f(x)| < +\infty.$$

For $f \in L^1(G)$, we may take $\mu(f) = \sum_{x \in S} f(x)$ and $\|f\| = \sum_{x \in S} |f(x)|$. For example, if $G = (\mathbf{Z}, +)$, then $L^1(\mathbf{Z})$ is the set of all doubly infinite sequences $\mathbf{a} = \{a_n\}_{n\in\mathbf{Z}}$ such that $\sum_{-\infty}^{+\infty} |a_n|$ converges, and $\mu(\mathbf{a}) = \sum_{-\infty}^{+\infty} a_n$.

If $\mathbf{a}, \mathbf{b} \in L^1(\mathbf{Z})$ then $\mathbf{a} * \mathbf{b} = \mathbf{c}$ where $c_n = \sum_{m=-\infty}^{+\infty} a_m b_{n-m}$ (the Cauchy product of Laurent series).

6. Let $G = (\mathbf{R}^{pos}, \cdot)$. If $f \in \mathcal{K}(G)$ then $\rho(f) = \int_0^\infty f(x)dx$ (Lebesgue integral) is a positive integral but it is not right invariant. In fact if $y \in \mathbf{R}^{pos}$, then

$$\begin{aligned}\rho(f^y) &= \int_0^\infty f(xy^{-1})dx \\ &= \int_0^\infty yf(z)dz \\ &= y\rho(f).\end{aligned}$$

A general procedure for finding the Haar integral in such cases is given by

Proposition 29. <u>Let</u> G <u>be a locally compact, Hausdorff group, and let</u> ρ <u>be a positive integral on</u> G. <u>Suppose that</u> $\rho(f^y) = g(y)\rho(f)$, <u>for all</u> $f \in \mathcal{K}(G)$ <u>and</u> $y \in G$, <u>where</u> g <u>is a function from</u> G <u>to</u> $\mathbf{R}^{pos}$. <u>Then the equation</u> $\mu(f) = \int_G \frac{f(x)}{g(x)} d\rho(x)$ <u>defines a right Haar integral</u> μ <u>on</u> G.

Proof. Since $f^{yz} = (f^y)^z$, and since $\rho(f) \neq 0$ for some $f \in \mathcal{K}(G)$ we see immediately that $g(yz) = g(y)g(z)$.

Hence
$$\begin{aligned}\mu(f^y) &= \int_G \frac{f(xy^{-1})}{g(x)} d\rho(x) \\ &= \int_G \frac{f(xy^{-1})}{g(xy^{-1})g(y)} d\rho(x) \\ &= \frac{1}{g(y)} \rho\left(\left(\frac{f}{g}\right)^y\right) \\ &= \rho\left(\frac{f}{g}\right) \\ &= \mu(f).\end{aligned}$$

Thus μ is a right Haar integral (cf. the relationship between left and right Haar integrals and modular functions).

Corollary. The Haar integral on $\mathbf{R}^{pos}$ is given by

$$\mu(f) = \int_0^\infty \frac{f(x)}{x}\, dx.$$

Alternatively we can obtain the Haar integral on $\mathbf{R}^{pos}$ from the isomorphism

$$G = (\mathbf{R}^{pos}, \cdot) \cong (\mathbf{R}, +) = H$$

given by $\log : G \to H$ and $\exp : H \to G$. If we transfer $f \in \mathcal{K}(G)$ to H by this isomorphism we get $f^* \in \mathcal{K}(H)$ where $f^*(x) = f(\exp(x))$. If ν is a Haar integral on H we get a Haar integral μ on G by putting

$$\mu(f) = \nu(f^*) = \int_{-\infty}^{+\infty} f(\exp(x))dx = \int_0^\infty \frac{f(y)}{y}\, dy.$$

7. Let $G = (\mathbf{C}^*, \cdot)$. We know that $\mathbf{C}^* \cong \mathbf{R}^{pos} \times S^1$, so taking polar coordinates (r, θ) we have

$$\mu(f) = \int_0^\infty \frac{dr}{r} \int_0^{2\pi} f(r, \theta)d\theta.$$

But $dxdy = rdrd\theta$, so

$$\mu(f) = \iint \frac{f(r, \theta)}{r^2}\, dxdy = \int_{\mathbf{R}^2} \frac{f(z)}{|z|^2}\, dxdy,$$

where $z = x + iy$.

8. The general linear groups. Let $G = GL_n(\mathbf{R})$. Then G is an open subset of $\mathbf{R}^{n^2}$ because it is defined by the inequality $\det(x_{ij}) \neq 0$. We write $T = (t_{ij})$ for a variable matrix, M for $\mathbf{R}^{n^2}$ and $\int_M dT$ for the n^2-dimensional Lebesgue integral.

Given $f \in \mathcal{K}(G)$, define $\rho(f) = \int_G f(T)dT = \int_M \bar{f}(T)dT$ where $\bar{f}$ is the extension of f which has value 0 outside G. Clearly ρ is a positive integral on G, and we shall apply Proposition 29 to convert it into a Haar integral. We therefore let $U \in G$ and consider $\rho(f^{U^{-1}}) = \int_G f(TU)dT$. For fixed U, the map $S \mapsto SU$ is a linear map from M to M and we need to find its Jacobian. The matrix of this linear map, with respect to the standard basis of M, has rows enumerated by pairs of integers (i, j) where i, j run from 1 to n. Similarly for the columns. The entry in the (i, j)-th row and (k, l)-th column of the matrix is

$$\frac{\partial((SU)_{ij})}{\partial(S_{kl})} = \delta_{ik} U_{lj} .$$

Arranging the basis of M in lexicographical order we therefore obtain the matrix

$$\begin{pmatrix} U' & 0 & \cdots & 0 \\ 0 & U' & \cdots & 0 \\ \vdots & \vdots & \ddots & \vdots \\ 0 & 0 & \cdots & U' \end{pmatrix}$$

where U' is the transpose of U. Thus the Jacobian of the transformation is $(\det U)^n$ and we have

$$\rho(f^{U^{-1}}) = |\det U|^{-n} \rho(f),$$

or, equivalently, $\rho(f^U) = |\det U|^n \rho(f)$. It follows, by Proposition 29, that the functional μ defined by

$$\mu(f) = \int_G \frac{f(T)}{|\det T|^n} \, dT$$

is a right Haar integral on G. By symmetry, μ is also a left Haar integral and therefore the general linear group is unimodular (though it is neither Abelian nor compact).

Exercise. Obtain a formula for the Haar integral on $GL_n(\mathbf{C})$.

9. <u>The affine groups.</u> Let $G = GA_n(\mathbf{R})$, the affine group on $\mathbf{R}^n$. Then G has a normal subgroup $N \cong \mathbf{R}^n$ which consists of all the translations of $\mathbf{R}^n$. The quotient group G/N is isomorphic with $GL_n(\mathbf{R})$ and G is a <u>split extension</u> of N by the general linear group. This means that G is isomorphic with the group of all pairs (t, T), where $t \in N$, $T \in GL_n(\mathbf{R})$ and multiplication in G is given by the formula

$$(t, T).(t', T') = (t + t'T, TT').$$

Here we identify N with $\mathbf{R}^n$, using additive notation, and $t'T$ denotes the transform of the vector t' by the matrix T. For the computation we shall identify G with this group of pairs.

We now apply Example 3(ii) to find the Haar integrals on G. If $f \in \mathcal{K}(G)$, we first compute $f^*(t', T') = \int_N f((t, I).(t', T'))d(t, I)$, where I is the identity of $GL_n(\mathbf{R})$. From the definition of the multiplication in G, we deduce that

$$\begin{aligned} f^*(t', T') &= \int_{\mathbf{R}^n} f(t + t', T')dt \\ &= \int_{\mathbf{R}^n} f(t, T')dt. \end{aligned}$$

This is a function of T' alone (i.e. f^* is constant on the cosets of N) and, according to Example 3(ii), we obtain a right Haar integral μ by integrating it with respect to the right Haar measure on $GL_n(\mathbf{R})$. This gives

$$\mu(f) = \int_{GL_n(\mathbf{R})} \int_{\mathbf{R}^n} \frac{f(t, T)}{|\det(T)|^n} \, dtdT,$$

by Example 8, the integrals being n-dimensional and n^2-dimensional Lebesgue integrals.

For the left Haar integral on $G = GA_n(\mathbf{R})$ we must first compute

$$\begin{aligned} f^\dagger(t', T') &= \int_N f((t', T').(t, I))d(t, I) \\ &= \int_{\mathbf{R}^n} f(t' + tT', T')dt. \end{aligned}$$

Substituting $s = tT'$, we obtain

$$\begin{aligned} f^\dagger(t', T') &= \frac{1}{|\det(T')|} \int_{\mathbf{R}^n} f(t' + s, T')ds \\ &= \frac{1}{|\det(T')|} \int_{\mathbf{R}^n} f(t, T')dt. \end{aligned}$$

The left Haar integral ν is therefore obtained by integrating this function of T' over GL_n, which gives $\nu(f) = \int_{\mathbf{R}^{n^2}} \int_{\mathbf{R}^n} \frac{f(t, T)}{|\det(T)|^{n+1}} \, dtdT.$

So $GA_n(\mathbf{R})$ is an example of a group which is <u>not</u> unimodular. Its right modular function is $\Delta(t, T) = |\det(T)|$.

10. <u>The orthogonal groups.</u> The computation of Haar integrals on orthogonal groups is harder. O_n^+ is not an open subset of $\mathbf{R}^{n^2}$, so the n^2-dimensional Lebesgue integral is useless; it always gives zero. What is needed is a parametrisation of O_n^+ by local Euclidean coordinates. In fact O_n^+, as a closed subgroup of GL_n, acquires a natural structure as a Lie group and one gets local parameters by taking Euclidean coordinates in its tangent spaces. There is then a standard method of constructing the Haar integral (see e.g. Chevalley [2], Chapter 5).

Alternatively, one can extend the theory of Haar integrals to locally compact homogeneous spaces and then use the embeddings $O_1^+ \to O_2^+ \to O_3^+ \to \dots$ to obtain the Haar integral on O_n^+ inductively, using the analogue of Example 3(ii). Here we need to know the invariant integrals on the coset spaces $O_n^+/O_{n-1}^+ \simeq S^{n-1}$. For example, on S^2 there is a familiar surface integral, invariant under the action of O_3^+, and this together with the Haar integral on $O_2^+ \cong S^1$ gives an easy formula for the Haar integral on O_3^+.

11. <u>Profinite groups.</u> Any profinite group G is a compact Hausdorff group, so has a unique normalized Haar integral. Any continuous function $f : G \to \mathbf{R}$ is uniformly continuous, so for $\varepsilon > 0$ we can find a neighbourhood V of e such that $|f(x) - f(y)| < \varepsilon$ whenever $xy^{-1} \in V$. Now V contains an open normal subgroup H of finite index n, say. On each coset of H, f varies by at most ε, so if we choose arbitrary representatives x_i $(i = 1, 2, \dots, n)$ for the cosets of H, the real number

$$\frac{1}{n} \sum_{i=1}^{n} f(x_i)$$

is determined to within ε. The integral $\mu(f)$ is then the limit of this sum as $H \to e$ in the sense that, given $\varepsilon > 0$, there is a neighbourhood V of e such that

$$\left|\mu(f) - \frac{1}{n} \sum f(x_i)\right| < \varepsilon$$

for all open normal subgroups $H \subseteq V$.

The corresponding Haar measure on G is determined by the fact that the measure of G is 1 and for an open normal subgroup H of index n the measure of each coset of H is 1/n. For example, if $G = \mathbf{Z}_p$, the group of p-adic integers, the subgroups $H_n = p^n G$ $(n = 0, 1, \ldots)$ form a fundamental system of neighbourhoods of 0, and if $H_m x$, $H_n y$ are any two cosets of these subgroups, either they are disjoint or one is contained in the other. It follows that every open set S is the union of a (countable) collection of disjoint cosets of the H_n and the measure of S is the sum of the known measures of these cosets. In fact, if r_n denotes the number of cosets of H_n contained in S, then the measure of S is $\lim_{n\to\infty} (r_n/p^n)$.

3. Representations of compact groups

A (complex) representation of a topological group G is a morphism of topological groups $\rho : G \to GL_n(\mathbf{C})$ for some n. The integer n is called the degree of the representation. If we consider $GL_n(\mathbf{C})$ as acting on an n-dimensional complex vector space V in the usual way (on the right), then a representation ρ of degree n gives an action of G on V by the rule $v. g = v\rho(g)$ for $v \in V$, $g \in G$. This makes V a right G-module over $\mathbf{C}$, that is, $v. (gh) = (v. g). h$, $v. e = v$ and, for each g, $v \mapsto v. g$ is $\mathbf{C}$-linear. Also the action is continuous in the sense that the map $\sigma : V \times G \to V$ defined by $(v, g) \mapsto v. g$ is continuous, where V has its canonical topology as a $\mathbf{C}$-space ($V \cong \mathbf{C}^n$, and the topology is independent of the basis). We call V a $\mathbf{C}G$-module for short.

Conversely, given an n-dimensional $\mathbf{C}G$-module V with continuous action $V \times G \to V$, one can obtain a representation $\rho : G \to GL_n(\mathbf{C})$ by choosing a basis $v_1, v_2, \ldots, v_n$ for V and letting $v_i. g = \sum_j \rho_{ij}(g) v_j$ for $i = 1, 2, \ldots, n$. The matrix $\rho(g) = (\rho_{ij}(g))$ is then a continuous function of g and satisfies $\rho(gh) = \rho(g)\rho(h)$. Two representations which come from isomorphic modules in this way (in particular from different bases of the same module) are called equivalent. Note: a morphism $\theta : V_1 \to V_2$ of $\mathbf{C}G$-modules is a $\mathbf{C}$-linear map satisfying $(x. g)\theta = (x\theta. g)$ for all $x \in V_1$, $g \in G$; it is automatically continuous.

For discrete groups G, we may omit all mention of continuity without altering the meaning of the above definitions, and we are then reduced to the purely algebraic concepts of classical representation theory. In this theory the group algebra $A = \mathbf{C}G$, whose elements are finite formal sums $\sum_i \alpha_i g_i$ with $\alpha_i \in \mathbf{C}$, $g_i \in G$, plays a fundamental role. This A is an associative algebra with respect to the multiplication induced by multiplication in G of the basis elements. Any G-module M over **C** becomes an A-module if we define

$$m(\sum \alpha_i g_i) = \sum \alpha_i (m g_i)$$

for $m \in M$, $\alpha_i \in \mathbf{C}$ and $g_i \in G$. Conversely, any A-module is obviously a G-module over **C** since G is embedded in A. Thus the terms 'G-module over **C**' and '**C**G-module' are interchangeable.

Now A itself has a right A-module structure induced by its multiplication, and in the special case when G is finite A is actually a finite-dimensional A-module. The corresponding representation is called the _regular representation_ of G and it is _faithful_, that is, it is an embedding of G in $GL_n(\mathbf{C})$ (where $n = |G|$).

We now state the basic theorems of the complex representation theory of a _finite_ group G. (See Serre [10] for direct proofs of these.)

Theorem A. G _has a faithful representation, so is isomorphic to a group of complex matrices._

Theorem B. _Every representation of_ G _is equivalent to a unitary one (that is, to a representation by unitary matrices)._

Theorem C. _Every representation of_ G _is completely reducible._

The definition of 'completely reducible' is as follows.

1. An A-module V is _simple_ or _irreducible_ if it is not $\{0\}$ and has no submodules other than $\{0\}$ and V.

2. An A-module is _semi-simple_ or _completely reducible_ if it is a (finite) direct sum of simple modules.

The terms 'irreducible' and 'completely reducible' are also applied to the corresponding representations. In terms of matrices, a representation is completely reducible if it is equivalent to a representation of the form

$$\rho(g) = \begin{pmatrix} \rho_1(g) & 0 & & 0 \\ 0 & \rho_2(g) & & 0 \\ & & \ddots & \\ 0 & 0 & & \rho_s(g) \end{pmatrix}$$

where each ρ_k is an irreducible representation.

Theorem D. All irreducible representations of G come from simple submodules of the group algebra $A = \mathbf{C}G$.

Theorem E. Let ρ run through one unitary representation from each equivalence class of irreducible representations. Then the functions $\rho_{ij} : G \to \mathbf{C}$ are orthogonal: if ρ, σ are two of these unitary representations, then

$$\frac{1}{|G|} \sum_{x \in G} \rho_{ij}(x) \overline{\sigma_{kl}(x)} = \delta_{\rho\sigma}\delta_{ik}\delta_{jl}(\deg \rho)^{-1}.$$

Theorem F. The functions ρ_{ij} in Theorem E form a basis for the space of all functions $G \to \mathbf{C}$. Equivalently, the elements $\sum_{x \in G} \rho_{ij}(x)x$ form a basis for the group algebra A.

It is natural to ask whether these important results can be proved for suitable topological groups and their (continuous) representations. We shall give an outline of the proofs of appropriate generalisations of all of them for an arbitrary compact Hausdorff group G.

The first thing to notice is that the group algebra $A = \mathbf{C}G$ of an infinite discrete group is an infinite-dimensional module and has, in fact, no finite-dimensional submodules except $\{0\}$ (exercise!); it is therefore useless for representation theory. However this group algebra can be regarded as the space of all functions $f : G \to \mathbf{C}$ with finite support; the element of $\mathbf{C}G$ corresponding to such an f is $\sum_{x \in G} f(x)x$. Since 'discrete and compact' means 'finite', and all functions on G are continuous, this space of functions is just $\mathcal{K}(G)$ (except that we are now looking at complex-valued functions). With this interpretation, the multiplication in A is convolution of functions (see §2, Example 4 above) and the action of G on A is right translation of functions. We may therefore enquire whether Theorems A-F can be proved for a special class of infinite topological groups using as 'group algebra' the convolution algebra $A = \mathcal{K}(G)$ consisting of all complex-valued continuous functions on G with compact support. Provided that G is locally compact and Hausdorff, we have a right Haar integral $\mu : \mathcal{K}(G) \to \mathbf{C}$ (integrate real and imaginary parts separately) and the recipe for generalisation is to replace $\sum_{x \in G} \phi(x)$ in the finite case by $\int_G \phi(x) d\mu(x)$.

For a compact group G this programme works very well and gives results which include Theorems A-F as special cases. The algebra A is now, of course, the algebra of all continuous functions $G \to \mathbf{C}$. It is a right G-module under right translation:

$$f^s(t) = f(ts^{-1}) \quad (f \in A; \ s, t \in G)$$

and we must study its finite-dimensional submodules. We use the normalized form μ of the right Haar integral and we recall that multiplication in A is given by

$$(f * g)(x) = \int_G f(xy^{-1}) g(y) d\mu(y).$$

(This formula, for complex-valued functions, gives a product with all the usual formal properties.)

We first observe that convolution and translation are compatible:

$$\begin{aligned}(f * g)^s(x) &= (f * g)(xs^{-1}) \\ &= \int_G f(xs^{-1}y^{-1}) g(y) d\mu(y) \\ &= \int_G f(xs^{-1}(ys^{-1})^{-1}) g(ys^{-1}) d\mu(y) \quad \text{(since } \mu \text{ is right invariant)} \\ &= \int_G f(xy^{-1}) g^s(y) d\mu(y) \\ &= (f * g^s)(x),\end{aligned}$$

that is, $(f * g)^s = f * g^s$,

It follows that the principal right ideal $f * A$ of A, for fixed f, is a G-submodule of A (though still not finite-dimensional). So let us fix $f \in A$ and write M for the submodule $f * A = \{f * a; \ a \in A\}$. We say that $g \in M$ is an <u>eigenfunction</u> for f if $f * g = cg$ for some complex number c. The eigenfunctions are the solutions of the integral equation

$$cg(x) = \int_G f(xy^{-1})g(y)d\mu(y).$$

For any such eigenfunction g, we have

$$f * g^s = (f * g)^s = (cg)^s = cg^s,$$

so g^s is also an eigenfunction for the same value of c. Hence the space $E_c = \{g \in M;\ f * g = cg\}$ is actually a submodule of M.

We shall assume the standard theory of integral equations, which is applicable without change to continuous functions on compact groups. (For details, see Chevalley [2], Chapter VI, or Smithies [11].) The basic result is that if $k \neq 0$ is a fixed symmetric, continuous function from $G \times G$ to $\mathbf{C}$, then the solutions g of the integral equation

$$cg(x) = \int_G k(x, y)g(y)d\mu(y),$$

for fixed c, form a finite-dimensional $\mathbf{C}$-space E_c; furthermore, the spaces E_c for $c \neq 0$ span topologically the space of all transforms

$$\int_G k(x, y)h(y)d\mu(y),$$

in the sense that any such transform is the sum of a uniformly convergent series $\sum g_i$, where each g_i lies in some $E_{c(i)}$ with $c(i) \neq 0$.

Applying this result with $k(x, y) = f(xy^{-1})$, $(f \in A)$, we see that if $f \in A$ is symmetric (i.e. $f(t^{-1}) = f(t)$) and $f \neq 0$ then the G-module $M = f * A$ is spanned topologically (in the above sense) by the finite-dimensional eigenspaces E_c, for $c \neq 0$, and each of these E_c is a submodule. There seems to be no shortage of finite-dimensional representations of G; but we must, of course, make sure that they are not all trivial. We shall show, indeed, that for every $t \neq e$ in G there is an $f \in A$ and a corresponding eigenspace E_c such that t acts non-trivially on E_c. For, if we are given $t \neq e$, we can find a neighbourhood U of e such that $t^{-1} \notin U^2$ and then, by Urysohn's lemma we can find a symmetric function $f \in A = \mathcal{K}(G)$ such that $f \geq 0$, $f(e) = 1$ and $f(x) = 0$

for $x \notin U$. For this f we have

$$(f * f)(e) = \int_G f(y^{-1})f(y)d\mu(y)$$
$$= \int_G \{f(y)\}^2 d\mu(y) > 0.$$

On the other hand

$$(f * f)^t(e) = (f * f)(t^{-1})$$
$$= \int_G f(t^{-1}y^{-1})f(y)d\mu(y)$$
$$= 0,$$

since if $f(t^{-1}y^{-1})f(y) \neq 0$ then $t^{-1}y^{-1} \in U$ and $y \in U$, whence $t^{-1} \in U^2$, which is false. Thus $(f * f)^t \neq f * f$ and so t acts non-trivially on $M = f * A$.

We remark that the action $\sigma : M \times G \to M$ given by $(h, t) \mapsto h^t$ is continuous when M is given the topology of uniform convergence (exercise!). Since the submodules E_c span linearly a dense subspace of M in this topology it follows that t must act non-trivially on some E_c. The topology on M induces on this finite-dimensional submodule E_c the canonical topology. Since G acts continuously on this module we get at last a continuous representation

$$\rho_t : G \to GL_n(\mathbf{C}), \quad (n = \dim E_c)$$

such that $t \notin \operatorname{Ker} \rho_t$. We write K_t for $\operatorname{Ker} \rho_t$ and M_t for the corresponding submodule of M (E_c above). Since $GL_n(\mathbf{C})$ is Hausdorff, K_t is closed in G; since $t \notin K_t$, the intersection of all the K_t is $\{e\}$; since G is compact, every neighbourhood V of e in G contains a finite intersection $K = K_{t(1)} \cap K_{t(2)} \cap \ldots \cap K_{t(r)}$. The direct sum of the corresponding modules

$$M_{t(1)} \oplus M_{t(2)} \oplus \ldots \oplus M_{t(r)}$$

is a finite-dimensional G-module with continuous action, giving rise to a continuous representation $\rho : G \to GL_n(\mathbf{C})$ whose kernel is precisely K. This proves:

Theorem A'. *If G is a compact Hausdorff group, then every neighbourhood of e contains a closed normal subgroup K such that G/K is isomorphic (in $\mathcal{TG}$) with a group of complex matrices.*

(*Note:* the continuous injection $G/K \to GL_m(\mathbf{C})$ is an isomorphism with the image group because G/K is compact and GL_m is Hausdorff.) An immediate corollary is

Theorem A". *Every compact Hausdorff group without small subgroups has a faithful representation as a group of complex matrices.*

(A group 'has small subgroups' if every neighbourhood of e contains a non-trivial subgroup; otherwise it is 'without small subgroups'.)

Clearly each of these theorems reduces to Theorem A if G is finite and discrete. With a little more work one can prove:

Theorem A‴. *Every compact Hausdorff group is the inverse limit of a system of compact matrix groups.*

Note: Since closed subgroups of $GL_m(\mathbf{C})$ inherit an analytic structure from it, every compact matrix group is a Lie group, and Theorem A‴ shows that compact Hausdorff groups can be 'approximated by Lie groups'. The corresponding statement for locally compact groups is true under some restrictions but is very much harder to prove. (See Kaplansky [4] and Montgomery and Zippin [6].)

The other results of representation theory mentioned earlier (Theorems B-F) are proved by methods which follow the classical proofs more closely. For example if $\rho : G \to GL_n(\mathbf{C})$ is any representation of a compact Hausdorff group G, let M be the corresponding module, and let $\langle , \rangle$ be any positive definite Hermitian form on M. Then the form $\langle , \rangle_G$ defined by

$$\langle u, v \rangle_G = \int_G \langle u.g, v.g \rangle \, d\mu(g)$$

is also positive definite Hermitian and has the extra virtue of being invariant under the action of G, i.e. $\langle u, v\rangle_G = \langle u.g, v.g\rangle_G$, by the left invariance of the Haar integral on G. (N.B., G is unimodular.) We may now choose an orthonormal basis of M with respect to $\langle , \rangle_G$, and the resulting representation, equivalent to ρ, will be unitary. This proves:

Theorem B'. Every representation of a compact Hausdorff group is equivalent to a unitary one.

Theorem C'. Every representation of a compact Hausdorff group is completely reducible.

Proof. This follows easily from Theorem B', or rather from its proof. We take a module M for the representation and construct an invariant positive definite Hermitian form on it. If M is not irreducible it has a proper submodule M_1. Let M_2 be the orthogonal complement of M_1. By the invariance of the form M_2 is also a submodule of M, and $M = M_1 \oplus M_2$. The result now follows by induction on dim M.

Theorem E'. If G is a compact Hausdorff group, and if ρ runs through a complete set of inequivalent irreducible, unitary representations of G, then the functions $\rho_{ij} : G \to \mathbf{C}$ are orthogonal:

$$\int_G \rho_{ij}(x)\, \overline{\sigma_{kl}(x)}\, d\mu(x) = \delta_{\rho\sigma}\delta_{ik}\delta_{jl}(\deg \rho)^{-1},$$

where μ is the normalized Haar integral on G.

Proof. This is essentially Schur's lemma. (Schur's lemma states that any non-zero morphism $\theta : M \to N$ between irreducible modules is an isomorphism. This is almost obvious since the image and kernel of θ are submodules of N and M respectively. Since the image of θ is not 0 it must be N, and since the kernel of θ is not M it must be 0.) Let ρ, σ be representations of degrees r, s arising from modules R, S with respect to fixed bases. Let $\alpha : R \to S$ be any linear map, and let A be the corresponding $r \times s$ matrix. By 'averaging over G' we obtain a module homomorphism $\alpha_0 : R \to S$, where for $r \in R$,

$$r\alpha_0 = \int_G ((rx)\alpha)x^{-1} d\mu(x).$$

By Schur's lemma, if ρ and σ are irreducible, α_0 is 0 or an isomorphism. Suppose first that ρ and σ are inequivalent irreducible representations. Then $\alpha_0 = 0$ for every choice of α. In matrix notation this says (assuming that the representations are unitary) that

$$\int_G \rho(A)\,\overline{\sigma(x)}\,d\mu(x) = 0$$

for all matrices A, where integration is performed component-wise. From this follows

$$\int_G \rho_{ij}(x)\,\overline{\sigma_{kl}(x)}\,d\mu(x) = 0 \quad \text{for all } i, j, k, l.$$

For the rest of the proof we may assume that $\rho = \sigma$ is an irreducible unitary representation of degree r coming from a fixed basis of the module $R = S$. If α is any linear map $R \to R$ then α_0 is a module endomorphism of R. Let λ be an eigenvalue of α_0 in **C**. Then $\alpha_0 - \lambda\iota$ is a singular module-endomorphism and so must be 0, by Schur's lemma. Thus $\alpha_0 = \lambda\iota$ for suitable $\lambda \in \mathbf{C}$. In matrix notation this says that for every $r \times r$ matrix A

$$\int_G \rho(x) A \overline{\rho(x)}\,d\mu(x) = \lambda I,$$

for some λ depending on A. Taking traces of both sides we find that trace $A = \lambda r$, so $\lambda = (\deg \rho)^{-1}$(trace A). The equations

$$\int_G \rho_{ij}(x)\,\overline{\rho_{kl}(x)}\,d\mu(x) = \delta_{ik}\delta_{jl}(\deg \rho)^{-1}$$

now follow by taking for A the single-entry matrices E_{jl}.

Theorem F'. (Peter-Weyl) Every continuous function from a compact Hausdorff group G to **C** can be uniformly approximated by linear combinations of the functions ρ_{ij} in Theorem E'.

Proof. First observe that if ρ and σ are equivalent representations, then each function σ_{kl} is a linear combination of the functions ρ_{ij} because, for some matrix T, $\sigma(x) = T\rho(x)T^{-1}$ for all $x \in G$. So the statement of the theorem is unaltered if we admit all the functions ρ_{ij} for all irreducible representations ρ. Indeed, since every representation (of finite degree) is equivalent to a diagonal sum of irreducible ones, it makes no difference if we admit the functions ρ_{ij} for all representations

sentations ρ. We therefore let $R \subseteq \mathcal{K}(G)$ be the closure (in the topology of uniform convergence) of the linear space generated by all the coordinate functions ρ_{ij} of all the representations ρ of G. It is enough to show that $R = \mathcal{K}(G)$.

Now, for any non-zero symmetric function $f \in \mathcal{K}(G)$, the G-module $M_f = f * \mathcal{K}(G)$ contains finite-dimensional submodules E_c which between them span a dense subspace (see the proof of Theorem A'). Let E_c be such a subspace, giving rise to a representation ρ with respect to a basis $g_1, g_2, \ldots, g_r$ of eigenfunctions. Then $g_i^x = \sum_j \rho_{ij}(x)g_j$, and so $g_i(x^{-1}) = \sum_j \rho_{ij}(x)g_j(e)$. This shows that $\hat{g}_i \in R$, where $\hat{g}_i(x) = g_i(x^{-1})$. In the obvious notation, this implies that $\hat{E}_c \subseteq R$ and hence that $\hat{M}_f \subseteq R$ for every symmetric f.

Finally, if h is any function in $\mathcal{K}(G)$, we can approximate it by elements of suitable modules M_f as follows. Let $\varepsilon > 0$ and choose a symmetric neighbourhood U of e in G such that

$$|h(x) - h(y)| < \varepsilon \text{ for } xy^{-1} \in U.$$

This is possible by Proposition 21. By Urysohn's lemma we can find a symmetric function $f > 0$ which vanishes outside U, and by normalization we may assume that $\int f(x)d\mu(x) = 1$. It is an easy exercise to show that $|(f * h)(x) - h(x)| < \varepsilon$ for all $x \in G$, and so h is a uniform limit of functions of the type $f * h \in M_f$. Since $\hat{M}_f \subseteq R$, this implies that $h \in R$ for every h, and therefore $R = \mathcal{K}(G)$.

Theorem D'. Every irreducible representation of the compact Hausdorff group G comes from a submodule of $\mathcal{K}(G)$.

Proof. This follows easily from the proof of the Peter-Weyl theorem and the orthogonality relations. For let ρ be an irreducible representation and suppose that it is not obtainable from a (finite-dimensional) submodule of $\mathcal{K}(G)$. Then its coordinates ρ_{ij} are orthogonal to all coordinates of all irreducible representations which do come from submodules of $\mathcal{K}(G)$. But these latter span a dense subspace of $\mathcal{K}(G)$ and it follows that the ρ_{ij} are orthogonal to all functions in $\mathcal{K}(G)$, in particular to themselves. So $\rho_{ij} = 0$, for all i, j, which is impossible for a representation.

4. Topics for further reading

(a) Invariant integrals on homogeneous spaces have been mentioned in connection with integration on orthogonal groups. This topic is treated in Nachbin [7], Chapter III.

(b) The representation theory of compact groups is carried much further in Adams [1] and is neatly done in coordinate-free fashion. I strongly recommend this book as an introduction to Lie groups.

(c) The character theory of locally compact Abelian groups can be found in Pontryagin [8] in a rather old-fashioned presentation (but none the less worth reading). The duality theory there presented is a cornerstone of abstract harmonic analysis. More accessible accounts can be found in Loomis [5] and Rudin [9].

(d) The structure theory of locally compact Abelian groups is closely related to (c), but can be carried through without characters or integration. This is done in Hewitt and Ross [3], Chapter 2. (This 2-volume work contains a super-abundance of information on topological groups, but because of the detail offered, it is sometimes difficult to read.) See also Montgomery and Zippin [6], p. 101.

(e) The solution of Hilbert's 5th problem. A full account of this has been written by two of the principals. Their book, Montgomery and Zippin [6], is very readable and contains a good treatment of much standard material. A slicker, but somewhat sketchy account is given in the second half of Kaplansky [4]. This little book is a good place to get the flavour of the subject, and the exposition carries one painlessly through the most formidable difficulties.

References

[1] J. F. Adams. Lectures on Lie groups. W. A. Benjamin, Inc., New York-Amsterdam (1969).

[2] C. Chevalley. Theory of Lie groups I. Princeton University Press (1946).

[3] E. Hewitt and K. A. Ross. Abstract harmonic analysis. 2 vols; Springer-Verlag, Berlin-Göttingen-Heidelberg (1963).

[4] I. Kaplansky. Lie algebras and locally compact groups. Chicago University Press (1971).

[5] L. H. Loomis. An introduction to abstract harmonic analysis. Van Nostrand Co., Princeton N. J. (1953).

[6] D. Montgomery and L. Zippin. Topological transformation groups. Interscience Publishers, Inc., New York (1955).

[7] L. Nachbin. The Haar integral. Van Nostrand Co., Princeton N. J. (1965).

[8] L. Pontryagin. Topological groups. Princeton University Press (1946).

[9] W. Rudin. Fourier analysis on groups. Interscience Publishers, Inc., New York (1962).

[10] J.-P. Serre. Représentations linéaires des groupes finis. Hermann, Paris (1967).

[11] F. Smithies. Integral equations. Cambridge University Press (1962).

[12] H. Widom. Lectures on measure and integration. Van Nostrand Mathematical Studies, No. 20 (1969).

Index

Index to numbered results